my **revision** notes

OCR

LEVEL 3 FREE STANDING MATHEMATICS QUALIFICATION: ADDITIONAL MATHS

2nd Edition

Michael Ling

Series editor:
Roger Porkess

HODDER EDUCATION
AN HACHETTE UK COMPANY

Although every effort has been made to ensure that website addresses are correct at time of going to press, Hodder Education cannot be held responsible for the content of any website mentioned in this book. It is sometimes possible to find a relocated web page by typing in the address of the home page for a website in the URL window of your browser.

Hachette UK's policy is to use papers that are natural, renewable and recyclable products and madefrom wood grown in well-managed forests and other controlled sources. The logging and manufacturing processes are expected to conform to the environmental regulations of the country of origin.

Orders: please contact Hachette UK Distribution, Hely Hutchinson Centre, Milton Road, Didcot, Oxfordshire, OX11 7HH. Telephone: +44 (0)1235 827827. Email education@hachette.co.uk. Lines are open from 9 a.m. to 5 p.m., Monday to Friday. You can also order through our website: www.hoddereducation.co.uk

ISBN: 978 1 5104 4960 2

First published in 2019 by
Hodder Education,
An Hachette UK Company
Carmelite House
50 Victoria Embankment
London EC4Y 0DZ
www.hoddereducation.co.uk

Impression number 10 9 8 7 6 5 4 3

Year 2023 2022

Cover photo © Rysuku/stock.adobe.com
Typeset in India by Aptara, Inc.
Printed by CPI Group (UK) Ltd, Croydon, CR0 4YY

A catalogue record for this title is available from the British Library.

Get the most from this book

Welcome to your Revision Guide for OCR Level 3 Free Standing Mathematics Qualification: Additional Mathematics. This book will provide you with reminders of the knowledge and skills you will be expected to demonstrate in the exam, with opportunities to check and practice those skills on exam-style questions. Additional hints and notes throughout help you to avoid common errors and provide a better understanding of what's needed in the exam.

Included with the purchase of this book is valuable online material that provides full worked solutions to all the 'Target your revision questions', 'Exam-style questions' and 'Review questions'. The online material is available at www.hoddereducation.co.uk/MRNOCRAdditionalMaths.

Features to help you succeed

Target your revision

Use these questions at the start of each section to focus your revision on the topics you find tricky. Short answers are at the back of the book, but use the full worked solutions online to check each step in your solution.

About this topic

At the start of each chapter, this provides a concise overview of its content.

Before you start, remember ...

A summary of the key things you need to know before you start the chapter.

Key facts

Check that you understand all the key facts in each subsection. These provide a useful checklist if you get stuck on a question.

Worked examples

Full worked examples show you what the examiner expects to see in order to ensure full marks in the exam. The examples cover a sample of the type of questions you can expect.

Hint

Expert tips are given throughout the book to help you do well in the exam.

Common mistakes

Your attention is drawn to typical mistakes students make, so you can avoid them.

Exam-style questions

For each topic, these provide typical questions you should expect to meet in the exam. Short answers are at the back of the book, and you can check your working using the full worked solutions online.

Review questions

After you have completed each section in the book, answer these questions for more practice. Short answers are at the back of the book, but the full worked solutions online allow you to check every line in your solution.

At the end of the book, you will find the following useful information.

Exam preparation

Includes hints and tips on revising for the OCR Free Standing Mathematics Qualification: Additional Maths exam, and information about the structure of the exam papers.

Make sure you know these formulae for your exam ...

Provides a succinct list of all the formulae you need to remember and the formulae that will be given to you in the exam.

Please note that the formula sheet as provided by the exam board for the exam may be subject to change.

During your exam

Includes key words to watch out for, common mistakes to avoid and tips if you get stuck on a question.

My revision planner

Section 1 Algebra

Section 2 Coordinate geometry in two variables

Section 3 Trigonometry

REVISED TESTED EXAM READY

Section 4 Selections

REVISED TESTED EXAM READY

Section 5 Powers and iteration

REVISED TESTED EXAM READY

Section 6 Calculus

Countdown to my exams

6–8 weeks to go

- Start by looking at the specification — make sure you know exactly what material you need to revise and the style of the examination. Use the revision planner on pages 4 and 5 to familiarise yourself with the topics.
- Organise your notes, making sure you have covered everything on the specification. The revision planner will help you to group your notes into topics.
- Work out a realistic revision plan that will allow you time for relaxation. Set aside days and times for all the subjects that you need to study, and stick to your timetable.
- Set yourself sensible targets. Break your revision down into focused sessions of around 40 minutes, divided by breaks. These Revision Notes organise the basic facts into short, memorable sections to make revising easier.

REVISED ☐

2–6 weeks to go

- Read through the relevant sections of this book and refer to the exam tips, exam summaries, typical mistakes and key terms. Tick off the topics as you feel confident about them. Highlight those topics you find difficult and look at them again in detail.
- Test your understanding of each topic by working through the 'Now test yourself' questions in the book. Look up the answers at the back of the book.
- Make a note of any problem areas as you revise, and ask your teacher to go over these in class.
- Look at past papers. They are one of the best ways to revise and practise your exam skills. Write or prepare planned answers to the exam practice questions provided in this book. Check your answers online
- Use the revision activities to try out different revision methods. For example, you can make notes using mind maps, spider diagrams or flash cards.
- Track your progress using the revision planner and give yourself a reward when you have achieved your target.

REVISED ☐

One week to go

- Try to fit in at least one more timed practice of an entire past paper and seek feedback from your teacher, comparing your work closely with the mark scheme.
- Check the revision planner to make sure you haven't missed out any topics. Brush up on any areas of difficulty by talking them over with a friend or getting help from your teacher.
- Attend any revision classes put on by your teacher. Remember, he or she is an expert at preparing people for examinations.

REVISED ☐

The day before the examination

- Flick through these Revision Notes for useful reminders, for example the exam tips, exam summaries, typical mistakes and key terms.
- Check the time and place of your examination.
- Make sure you have everything you need — extra pens and pencils, tissues, a watch, bottled water, sweets.
- Allow some time to relax and have an early night to ensure you are fresh and alert for the examinations.

REVISED ☐

My exam

Date: ....

Time: ....

Location: ....

Section 1 Algebra

Target your revision (Chapters 1–4)

Try answering each question below. If you get stuck, follow the page reference underneath to revise that topic.

1 **Manipulating algebraic expressions**
Simplify $2(x-3y)+3(y+1)+x$.
(see page 3)

2 **Manipulating expressions involving square roots**
Simplify $\sqrt{18}+\sqrt{32}$.
(see page 5)

3 **Rationalise the denominator of a fraction involving surds**
Simplify $\frac{3}{2-\sqrt{3}}$.
(see page 6)

4 **Work with polynomials**
You are given that $f(x)=x^3-x^2+x-1$ and $g(x)=3x+1$.
Find
(i) $f(x)+g(x)$,
(ii) $f(x)g(x)$.
(see page 8)

5 **Review of solving linear equations**
Solve $3(x-2)-1=1+x$.
(see page 10)

6 **Solving quadratic equations that can be factorised**
Solve $2x^2-x-3=0$.
(see page 11)

7 **Solving quadratic equations that cannot be factorised**
Solve $2x^2+x-5=0$.
(see page 11)

8 **Checking for roots of a quadratic equation**
Show that the equation $x^2-x+7=0$ has no real roots.
(see page 11)

9 **Complete the square for a quadratic function**
By completing the square, show that $n^2-6n+15$ is positive for all values of n.
(see page 11)

10 **Quadratic graphs**
Sketch the curve $y=x^2-2x-6$.
From your graph write down the roots of the equation $x^2-2x-6=0$. Give your answers to 1 decimal place.
(see page 11)

11 **The factor theorem**
Given that $f(x)=3x^2+4x-4$, show that $f(x)$ has a factor $(x+2)$.
(see page 13)

12 **Solving cubic equations using the factor theorem**
Given that $f(x)=x^3-2x^2-5x+6$, solve $f(x)=0$.
(see page 13)

13 **Review of simultaneous equations when both are linear**
Solve the simultaneous equations $2x+5y=19$ and $6x-y=9$.
(see page 15)

14 **Solving simultaneous equations when one in linear and the other quadratic**
Solve the simultaneous equations $y=x^2-3x+5$ and $y=5x-7$.
(see page 15)

15 **Setting up equations**
Last year Paul was twice as old as John. This year the sum of their ages is 77. Find the ages of Paul and of John this year.
(see page 17)

16 **Linear inequalities**
Solve $3<2x-1<19$.
(see page 19)

17 **Quadratic inequalities**
Plot the curve $y=x^2+x-6$ and hence find the range of values for which $x^2+x-6\leqslant 0$.
(see page 19)

18 **1st order recurrence relationships**
Term to term rule
Given that $u_n=2u_{n-1}+1$ with $u_1=1$, find the first 5 terms of the sequence.
(see page 22)

19 **1st order recurrence relationships**
Position to term rule
Given that $u_n = n^2 + 1$, find the first 5 terms of the sequence.

(see page 22)

20 **2nd order recurrence relationships**
Given that $u_{n+2} = 2u_{n+1} + u_n$ with $u_1 = 1$ and $u_2 = 2$ find the first 5 terms of the sequence.

(see page 24)

21 **Modelling using recurrence relationships**
Developers believe that their house building programme in a particular small town will result in 5% growth of the population per year. This year the population numbers 10 000. Find the predicted population in 10 years' time.

(see page 23)

Short answers on page 119
Full worked solutions online

CHECKED ANSWERS

Chapter 1 Algebraic manipulation

About this topic

Algebra is fundamental to higher level mathematics and underpins all of the work you will do in this course. Many of the questions that you will meet will expect you to manipulate algebraic expressions on the way to a required result.

You will work with square roots and practice how to manipulate them. Sometimes you will meet fractions where the denominator (the bottom line of a fraction) contains a square root. This can make them awkward to work with. However, you can simplify such fractions by 'rationalising the denominator' which is a process that ends up with square roots only in the numerator (the top line of the fraction).

Before you start, remember ...

- how to expand brackets
- how to add and subtract algebraic terms
- how to find the lowest common multiple (LCM) of two numbers or terms
- how to multiply algebraic terms
- how to manipulate square roots.

1.1 Manipulating algebraic expressions

REVISED

Key facts

1 **Collecting 'like terms'**

'Like terms' should be collected together.

For example: $a+2b+3a-b=(a+3a)+(2b-b)=4a+b$

2 **Expanding brackets**

Every term inside the bracket must be multiplied by the term outside.

For example: $2(x+3)=2x+6$

3 **Multiplying fractions**

Look for common factors which may be numbers or letters or even expressions.

For example: $\frac{2x^2}{y}\times\frac{y}{4x}$

$=\frac{2\times x\times x}{4\times x}=\frac{2x}{2\times 2}$

$=\frac{x}{2}$

4 **Adding fractions**

As with arithmetic you need to find a common denominator which is the lowest common multiple of the denominators (LCM).

For example: $\frac{a}{b}+\frac{c}{d}=\frac{ad+bc}{bd}$

'Like terms' are those that can be added together.

y can be cancelled,

then x,

then 2.

The LCM is the product of the denominators, that is bd.

Worked example

Collecting terms and factorisation

1 Tidying up an expression will usually involve collecting like terms.

Simplify $3xy-2y^2+xy-4y^2$.

Solution

$3xy-2y^2+xy-4y^2$

$=4xy-6y^2$

$=2y(2x-3y)$

The terms in xy are 'like' terms, as are the terms in y^2.

Then factorise.

Worked example

Multiplying out and collecting like terms

2 Like terms can often be seen when brackets are removed.

Simplify $5(x-2)-3(2-x)$.

Solution

Multiply out the brackets taking care of the signs.

$5(x-2)-3(2-x)$

$=5x-10-6+3x$

$=8x-16$

$=8(x-2)$

Then collect like terms.

In this example a factorisation is now possible.

Common mistake: If you are subtracting a bracket in which there is a negative sign then when you remove the bracket that term is positive.

Worked example

Multiplying fractions

3 When algebraic fractions are multiplied together, look for common factors in the numerator and denominator.

Simplify $\frac{2x^2}{3y}\times\frac{9y^2}{8xy}$.

Solution

$\frac{2x^2}{3y}\times\frac{9y^2}{8xy}$

$=\frac{2\times 9\times x^2\times y^2}{8\times 3\times x\times y^2}$

$=\frac{1\times 3\times x\times 1}{4\times 1\times 1\times 1}=\frac{3x}{4}$

In the re-written form, you can see that there are common factors in the numerator and denominator of 2, 3, x and y^2.

Worked example

Adding fractions

4 The principle for adding algebraic fractions is the same as for arithmetic fractions.

Write $\frac{2}{x+1}-\frac{3}{x+2}$ as a single fraction in its simplest form.

Solution

$$\frac{2}{x+1}-\frac{3}{x+2}$$

$$=\frac{2}{x+1}\times\frac{x+2}{x+2}-\frac{3}{x+2}\times\frac{x+1}{x+1}$$

In this case the LCM is $(x+1)(x+2)$.

$$=\frac{2(x+2)-3(x+1)}{(x+1)(x+2)}$$

Then collect like terms. Don't forget to multiply out the numerator of both fractions.

$$=\frac{-x+1}{(x+1)(x+2)}$$

Exam-style question

TESTED

Write $\frac{2}{x}+\frac{x-3}{5}$ as a single fraction in its simplest form.

Short answer on page 119

Full worked solution online

CHECKED ANSWER

1.2 Manipulating expressions involving square roots

REVISED

Key facts

1 **Irrational numbers**

A number which cannot be expressed as a simple fraction (or integer) is described as irrational.

For example: 3π, $3\sqrt{2}$, $\frac{2+\sqrt{5}}{7}$

An irrational square root is often called a surd.

2 **Square roots**

A square root of an integer is either itself an integer (e.g. $\sqrt{16}=4$) or it is irrational (e.g. $\sqrt{5}$).

Square roots can often be simplified.

For example: $\sqrt{18}=\sqrt{9\times 2}=\sqrt{9}\sqrt{2}=3\sqrt{2}$

A number whose square root is an integer is called a perfect square. So, for example 16 is a perfect square since $\sqrt{16}=4$.

3 **Rationalising the denominator of fractions involving square roots**

In the fraction $\frac{3}{2+\sqrt{2}}$ the denominator is irrational. By multiplying top and bottom of the fraction by $2-\sqrt{2}$ the denominator becomes rational since

$$\left(2-\sqrt{2}\right)\left(2+\sqrt{2}\right)=4+2\sqrt{2}-2\sqrt{2}-\sqrt{2}\sqrt{2}$$
$$=4-2=2$$

Note the difference of squares.

This process is called 'rationalising the denominator'.

Worked examples

Manipulating numeric expressions involving square roots

Square roots are treated in exactly the same way as ordinary numbers or letters in algebra.

1 Find $\left(\sqrt{3}-\sqrt{2}\right)\left(\sqrt{3}+\sqrt{2}\right)$.

Solution

$\left(\sqrt{3}-\sqrt{2}\right)\left(\sqrt{3}+\sqrt{2}\right)$

$=\sqrt{3}\times\sqrt{3}+\sqrt{3}\times\sqrt{2}-\sqrt{2}\times\sqrt{3}-\sqrt{2}\times\sqrt{2}$

$=3+\sqrt{6}-\sqrt{6}-2=1$

2 Simplify $\sqrt{54}-\sqrt{24}$.

Solution

$\sqrt{54}-\sqrt{24}=\sqrt{9\times6}-\sqrt{4\times6}$

$=3\sqrt{6}-2\sqrt{6}=\sqrt{6}$

Hint: Sometimes you will be asked for an approximate answer (usually to 3 significant figures) and at other times you will be asked for an exact value. Not all calculators will be able to do this for you.

Multiply out in the usual way $\sqrt{3}\times\sqrt{3}=3$.

Note that $\sqrt{54}$ and $\sqrt{24}$ are not in their simplest form. When you extract the factors that are perfect squares they become $3\sqrt{6}$ and $2\sqrt{6}$ and so are like terms.

Worked example

Manipulating algebraic expressions involving square roots

3 Remember that algebra is simply 'general arithmetic'.

Simplify $\sqrt{8a^3}+\sqrt{32a^5}$.

Solution

$\sqrt{8a^3}+\sqrt{32a^5}$

$=\sqrt{4a^2\times2a}+\sqrt{16a^4\times2a}$

$=2a\sqrt{2a}+4a^2\sqrt{2a}$

$=2a(1+2a)\sqrt{2a}$

Extract from the square root factors which are perfect squares. These are, in this example, 4 and 16, and also a^2 and a^4.

Worked example

Rationalising the denominator of a fraction

4 Simple fractions will terminate $\left(\text{e.g. } \frac{3}{5}\right)$ or recur $\left(\text{e.g. } \frac{3}{7}\right)$ when turned into decimals but this is not true for irrational numbers such as π and some square roots. Their decimals go on for ever and have no recurring pattern.

'Rationalising' means making the denominator a rational number or an integer.

Simplify $\frac{5}{5-\sqrt{2}}$.

Solution

$$\frac{5}{5-\sqrt{2}}=\frac{5}{5-\sqrt{2}}\times\frac{5+\sqrt{2}}{5+\sqrt{2}}=\frac{25+5\sqrt{2}}{25-2}=\frac{5}{23}\left(5+\sqrt{2}\right)$$

When leaving an arithmetic or algebraic term as an exact answer it is usual to express the term in the form $a\sqrt{b}$ or $a+b\sqrt{c}$ where a, b and c are rational numbers.

Terms such as $\frac{a}{\sqrt{2}}$ or $\frac{1}{1+\sqrt{a}}$ are not in this form so they need to be manipulated so that they are. The process usually involves multiplying the top and bottom of such a fraction by a number that makes the denominator rational.

Note the form of the expression you use to rationalise the denominator.

Exam-style question

TESTED

In this question you must show detailed reasoning.

Write $\frac{4-\sqrt{2}}{4+\sqrt{2}}$ in the form $a+b\sqrt{c}$ where a, b and c are numbers to be determined.

Short answer on page 119

Full worked solution online

CHECKED ANSWER

This is a common demand to indicate that you must show all working and not just give an answer which has been given by your calculator.

For instance, a numeric answer is 0.478 correct to three significant figures, but this is not what the question asks for and you will be given no marks for it. Even the answer given in the correct form may receive no marks without the working.

Chapter 2 Polynomials

About this topic

A polynomial is a function with a number of terms of the form ax^n added or subtracted, where n is a positive integer (or zero).

e.g. $f(x) = x^3 + x^2 - 2x + 4$, $f(x) = x^2 - 3x + 7$, $f(x) = 3x - 1$.

The order of a polynomial is the highest power of x.

A polynomial of order 3 is a cubic. One of order 2 is a quadratic and one of order 1 is a linear function

A polynomial may also include a constant term (where $n = 0$) but cannot contain terms such as $\sqrt{x}$ or $\frac{1}{x}$ as the powers of x are $\frac{1}{2}$ or, 1 and -1 which are not positive integers or zero.

Polynomials often occur in mathematics and so you need to be able to handle them with ease. Quadratics are particularly common

This chapter also includes solving polynomial equations involving linear, quadratic and cubic functions.

Before you start, remember ...

- how to add and subtract algebraic expressions with some like terms and some unlike terms from GCSE
- how to multiply out brackets
- how to factorise linear and quadratic expressions.

2.1 Polynomials

REVISED

Key facts

1 **Notation**

If a polynomial is denoted f(x) then f(a) is the value of the polynomial when $x = a$.

e.g. For $f(x) = x^3 + 2x - 5$, $f(2) = 2^3 + 2 \times 2 - 5 = 7$.

2 **Adding and subtracting polynomials**

Polynomials are added or subtracted in the same way as algebraic expressions. 'Like terms' are those with the same power.

3 **Multiplying polynomials**

When two polynomials are multiplied together then all the terms in one polynomial are multiplied by the terms in the other polynomial and then any like terms added.

4 **Dividing polynomials**

Division should be set out like an arithmetic long division where, instead of dealing in powers of 10, you are dealing with powers of x. When one polynomial is divided into another the quotient is the number of times it goes; there may be a remainder. Both the quotient and the remainder are polynomials or numbers.

Worked example

Multiplying polynomials

1 Multiply x^2+x-3 by $x+2$.

Solution

Method 1 Expansion

$$(x+2)(x^2+x-3)=x(x^2+x-3)+2(x^2+x-3)$$
$$=x^3+x^2-3x+2x^2+2x-6$$
$$=x^3+3x^2-x-6$$

Method 2 Long multiplication

$$(x+2)(x^2+x-3)=x^3+x^2-3x$$
$$+2x^2+2x-6$$
$$=x^3+3x^2-x-6$$

Multiply all three terms of the quadratic by x.

And then by 2.

Add like terms.

Worked example

Dividing polynomials

2 Divide x^3+2x^2-3x+2 by $x-3$

Solution

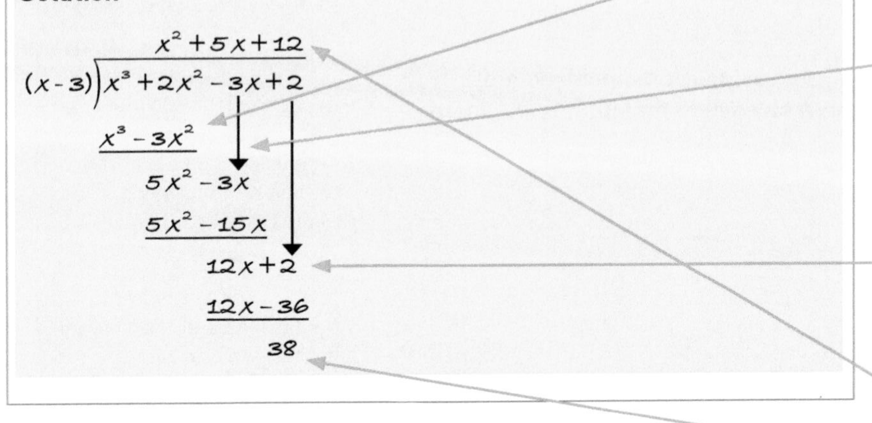

Lay out your work as shown. Multiply $(x-3)$ by x^2 and place under the first two terms. Subtract and put x^2 on the top line.

Bring down the next term so that you have $5x^2-3x$

Multiply $x-3$ by $5x$ and place under as shown. Place $+5x$ on the top line.

Repeat the process, leaving a remainder of 38. The quotient, the number of times that $x-3$ goes into the cubic function, is shown on the top line.

In this example it is $x^2+5x+12$.

The remainder is 38.

Exam-style question

TESTED

(i) Simplify $x^3+4x-3+2(x-2)(x+1)$.
(ii) Find, by long division, the remainder when x^3+2x^2+2x+5 is divided by $x+1$.

Short answer on page 119

Full worked solution online

CHECKED ANSWERS

2.2 Review of solving linear equations

REVISED

Key facts

1. **Solving linear equations**
   Manipulate both sides of the equation in the same way so as to get the terms in the unknown on one side of the equation.
   These steps can be carried out in any order depending on the complexity of the equation being solved.
2. **Solving equations graphically**
   To solve $f(x) = 0$, draw the graph of $y = f(x)$. The roots of the equation are where it crosses the x-axis.

The general principles are

- clear fractions
- multiply out brackets
- collect terms in the unknown on one side and the numbers on the other.

A value of x where the curve crosses the x-axis is called a root. A quadratic curve may have 2 roots, and a cubic up to 3 roots.

The 'solution' of an equation is the set of all its roots. However, in everyday use the word is often used to mean a 'root'.

Worked examples

Solving linear equations

In this example the equation has brackets but no fractions.

1 Solve the equation $2(x-3)=10-3(7-x)$.

Solution

$$2(x-3)=10-3(7-x)$$

Multiply out the brackets. Take care over the signs!

$$\Rightarrow \quad 2x-6=10-21+3x$$

$$\Rightarrow \quad 21-10-6=3x-2x$$

Collect like terms. It does not matter which side you use for the variable – in this case the right-hand side ensures that the terms remain positive and so sign errors are less likely.

$$\Rightarrow \quad 5=x$$

So $x=5$

Notice that the answer has the form $x = \ldots$ Of course, $x = 5$ is the same as $5 = x$.

In the next example, the equation involves brackets and fractions. In questions like this there can be several ways to manipulate the terms.

2 Solve the following equation.

$$\frac{2(x-3)}{5}+x=\frac{2-3x}{3}$$

Solution

$$\frac{2(x-3)}{5}+x=\frac{2-3x}{3}$$

It is a good idea to start by clearing the fractions. It simplifies the rest of the working and so you are less likely to make an error.

$$\Rightarrow \quad \frac{2(x-3)\times 15}{5}+15x=\frac{(2-3x)\times 15}{3}$$

The lowest common denominator of the fractions is 15. So to clear the fractions in this case, multiply through by 15.

$$\Rightarrow \quad 6(x-3)+15x=10-15x$$

Clear fractions.

$$\Rightarrow \quad 6x-18+15x=10-15x$$

Collect like terms.

$$\Rightarrow \quad 36x=28$$

$$\Rightarrow \quad x=\frac{7}{9}$$

Give your answer as a fraction in its lowest terms.

Common mistake: When you multiply throughout by 15, every term must be multiplied including, in this case, the x.

Exam-style question

TESTED

Solve the following equation.

$$\frac{2(x-1)}{3} - 2 = \frac{3x+1}{4}$$

Short answer on page 120
Full worked solution online

CHECKED ANSWER

2.3 Solving quadratic equations

REVISED

Key facts

1 **Solving quadratic equations that factorise**
The solution of the factorised equation $(x-a)(x-b)=0$ is $x=a$ or $x=b$.

2 **Solving quadratic equations that do not factorise**
The formula for solving the quadratic equation $ax^2+bx+c=0$ is $x=\frac{-b\pm\sqrt{b^2-4ac}}{2a}$.

3 **Checking for roots of a quadratic equation**
The expression b^2-4ac in the formula is called the discriminant.
If $b^2-4ac>0$ there are two distinct roots.
If $b^2-4ac=0$ the roots are coincident.
If $b^2-4ac<0$ there are no real roots.

4 **Completing the square**
The equation $ax^2+bx+c=0$ is written $a\left(\left(x+\frac{b}{2a}\right)^2-\left(\frac{b}{2a}\right)^2\right)+c=0$

which leads to

$$a\left(x+\frac{b}{2a}\right)^2=a\left(\frac{b}{2a}\right)^2-c$$

$$\left(x+\frac{b}{2a}\right)^2=\left(\frac{b}{2a}\right)^2-\frac{c}{a}$$

$$\Rightarrow x+\frac{b}{2a}=\pm\sqrt{\left(\frac{b}{2a}\right)^2-\frac{c}{a}}=\pm\sqrt{\frac{b^2-4ac}{4a^2}}$$

5 **Quadratic graphs**
The graph of the curve $y=ax^2+bx+c$ is a parabola.
If $a>0$ then the curve is the 'U-shaped' and is symmetric about the line $x=-\frac{b}{2a}$ with a minimum point.
If $a<0$ then the curve is '∩-shaped' and is symmetric about the line $x=-\frac{b}{2a}$ with a maximum point.
If the curve does not cut the x-axis, then there are no roots, corresponding to $b^2-4ac<0$.

In this course, a negative number is taken to have no square root. In more advanced mathematics, it is called an imaginary number and the square root of −1 is denoted by the letter i. This contrasts with real numbers, which are all those that you have worked with so far. Using imaginary numbers it is possible to find two roots for every quadratic equation.

Worked examples

Solving quadratic equations

In the first example the equation can be factorised.

1 Solve the following equation.

$2x^2 - 7x - 4 = 0$

Solution

$$2x^2 - 7x - 4 = 0$$
$$\Rightarrow \quad 2x^2 - 8x + x - 4 = 0$$
$$\Rightarrow \quad 2x(x-4) + 1(x-4) = 0$$
$$\Rightarrow \quad (2x+1)(x-4) = 0$$
$$\Rightarrow \quad x = -\frac{1}{2} \text{ or } x = 4$$

Find 2 numbers that multiply to give $2 \times -4 = -8$ and add to give -7. These numbers are -8 and 1.

Factorise the first two terms and factorise the last two terms, giving a common factor.

Factorise completely.

Don't forget to give your answers!

In the next example the equation cannot be factorised.

2 Solve the following equation.

$2x^2 - 4x - 7 = 0$

Solution

$2x^2 - 4x - 7 = 0$

In the quadratic formula $x = \frac{-b \pm \sqrt{b^2 - 4ac}}{2a}$

$a = 2$, $b = -4$ and $c = -7$

$$\Rightarrow \quad x = \frac{4 \pm \sqrt{16 + 56}}{4} = \frac{4 \pm \sqrt{72}}{4}$$
$$\Rightarrow \quad x = 1 \pm \frac{3}{2}\sqrt{2}$$
$$\Rightarrow \quad x = 3.12 \text{ or } x = -1.12 \text{ to 3 s.f.}$$

There are no numbers that multiply to give $2 \times -7 = -14$ and add to give -4 so the formula must be used.

If you are asked to give the answer exactly then this is the answer you should give.

If not, then use your calculator. The answer will then need to be rounded. Unless you are told otherwise give them to 3 significant figures.

Worked example

Solving by completing the square

3 Express $x^2 + 2x - 5 = 0$ in the form $(x+p)^2 = q$ and hence solve the equation.

Solution

$$x^2 + 2x - 5 = 0$$
$$\Rightarrow \quad x^2 + 2x = 5$$
$$\Rightarrow \quad x^2 + 2x + 1 = 5 + 1$$
$$\Rightarrow \quad (x+1)^2 = 6$$
$$\Rightarrow \quad x + 1 = \pm\sqrt{6}$$
$$\Rightarrow \quad x = -1 \pm \sqrt{6}$$
$$x = 1.45 \text{ or } -3.45 \text{ to 3 s.f.}$$

It is easiest to place the number on the right-hand side of the equation.

Add a number to both sides so that the left-hand side is a perfect square.

Take square roots of both sides.

This is the answer you should give if an exact answer is required.

This is given to 3 significant figures.

Worked example

Finding the minimum value of a quadratic function

4 Find the minimum value of $f(x)=x^2-4x+7$ by completing the square.

An alternative method for finding the minimum involves the use of calculus. This is covered in Chapter 14.

Solution

$$f(x)=x^2-4x+7$$
$$=(x^2-4x+4)+3$$
$$=(x-2)^2+3$$

Since the minimum value of the squared term is 0, the minimum value of f(x) is 3.

The minimum occurs when $x=2$. The equation of the line of symmetry of $y=f(x)$ is $x=2$.

Exam-style question

TESTED

(i) Express the quadratic expression x^2-3x-1 in the form $(x-p)^2-q$ where p and q are to be determined.

(ii) Hence solve the equation $x^2-3x-1=0$.

Short answer on page 120
Full worked solution online

CHECKED ANSWERS

2.4 Solving cubic equations

REVISED

Key facts

1 **Roots of a cubic equation**

A cubic equation has 3 roots, 2 of which may be coincident, or else 1 root.

2 **Solving a cubic equation**

If one or more of the roots is an integer then the factor theorem can be used.
Otherwise you need to use numerical methods to find the roots.

3 **The factor theorem**

If $(x-a)$ is a factor of $f(x)$ then $f(a)=0$.

Numerical methods are covered in Chapter 13.

For example: for $f(x)=x^3-2x+1$, $f(1)=1-2+1=0$. So $x=1$ is a root of the equation $x^3-2x+1=0$.

Worked examples

Using the factor theorem to solve a cubic equation

In this example, the equation has three roots, all of which are integers.

1 Solve the equation $f(x)=x^3+4x^2+x-6=0$.

Solution

The factors of -6 are 1, 2, 3, -1, -2 and -3.

By trial $f(1) = 0$.

So $x = 1$ is one of the roots.

Divide $f(x)$ by $(x - 1)$

Gives $f(x)=(x-1)(x^2+5x+6)=0$

$\Rightarrow \quad f(x)=(x-1)(x+3)(x+2)=0$

$\Rightarrow$ The roots are $x = 1, -2, -3$.

Try these numbers until you find one for which $f(x) = 0$.

Note that the factor theorem is an efficient way to find a root. However, it does not find the quotient which you may need to solve the cubic equation.

In this case you could have kept on trying the factors of -6 and you would have found the other two roots, but this will not always be the case as they need not be integers.

In the next example, the equation only has one real root.

2 Solve the equation $f(x)= x^3-3x^2+5x-6=0$.

Solution

The factors of -6 are 1, 2, 3, -1, -2 and -3.

By trial $f(2) = 8 - 12 + 10 - 6 = 0$.

So $x = 2$ is a root.

$(x^3-3x^2+5x-6)\div(x-2)=x^2-x+3$

So the equation is

$f(x)=(x-2)(x^2-x+3)=0$

In $x^2-x+3=0$, $b^2-4ac<0$ and so there are no real roots.

So $x = 2$ is the only root.

Exam-style question

TESTED

The function $f(x)$ is defined by $f(x)=x^3-5x^2+2x+8$.

(i) Find the value of $f(-1)$.

(ii) Solve the equation $f(x) = 0$.

Short answer on page 120

Full worked solution online

CHECKED ANSWERS

Chapter 3 Applications of equations and inequalities in one variable

About this topic

You learn algebra to solve problems! A typical problem will be given in words and you will need to set up the equation and solve it. This chapter covers ways of solving linear, quadratic and cubic equations. You will also meet simultaneous equations – two equations that relate two variables.

Linear and quadratic inequalities are also important in the solution of problems.

Before you start, remember ...

- the factor theorem.

3.1 Review of simultaneous equations

REVISED

Key facts

1 **Solving simultaneous equations when both are linear**

The two equations can be solved by elimination or by substitution.

- Elimination means manipulating the equations so that the coefficients of one variable are the same. Then the two equations are added or subtracted to leave a linear equation in one unknown.
- Substitution means making one of the variables the subject of one of the equations and substituting in the second equation.

2 **Solving simultaneous equations when one is linear and the other quadratic**

The equations need to be solved by substitution. Make one variable the subject of the linear equation and substitute in the quadratic equation.

Worked examples

Solving simultaneous equations

In this example both equations are linear.

1 Solve the following simultaneous equations

(i) by elimination and

(ii) by substitution.

$2x+5y=19$, $3x-y=3$

Solution

(i) Method 1 – elimination

$2x+5y=19$ ①

$3x-y=3$ ②

②×5: $15x-5y=15$ ③

$2x+5y=19$ ①

Add $17x = 34$ ①+③

$\Rightarrow x=2$

Make one coefficient the same or the negative of the other.

Add the two equations since in this case one of the coefficients is negative. Then solve for x.

Substitute for x into ①

$4+5y=19$

$\Rightarrow\ 5y=15$

$\Rightarrow\ y=3$

So the solution is $x=2$, $y=3$.

(ii) Method 2 – substitution

$2x+5y=19$ ①

$3x-y=3$ ②

② $y=3x-3$

In ①: $2x+5(3x-3)=19$

$\Rightarrow\ 17x-15=19$

$\Rightarrow\ 17x=34$

$\Rightarrow\ x=2$ and so $y=3$

Then substitute into one of the equations to find y.

Make one variable the subject of one equation and then substitute.

In the next example one equation is linear and the other is quadratic.

2 Find the points of intersection of the line $2x-y=11$ and the curve $y=x^2-5x-5$.

Solution

$2x-y=11$

$\Rightarrow\ y=2x-11$

$\Rightarrow\ 2x-11=x^2-5x-5$

$\Rightarrow\ x^2-7x+6=0$

$\Rightarrow (x-6)(x-1)=0$

So either $x=6$ giving $12-y=11 \Rightarrow y=1$

or $x=1$ giving $2-y=11 \Rightarrow y=-9$.

So the coordinates of the points of intersection are $(6, 1)$ and $(1, -9)$.

Make y the subject of the linear equation and substitute to give a quadratic equation in x.

Solve for x. Remember that there will be two values.

Substitute each value into the linear equation to find the associated value of y.

Remember to pair off the values correctly.

Common mistake: The question asks for the points of intersection, so leaving the answer with two values of x and two values of y is not a complete answer.

Exam-style question

TESTED

Solve simultaneously the equations $y=x+6$ and $y=x^2-x+3$.

Short answer on page 120

Full worked solution online

CHECKED ANSWER

3.2 Setting up equations

REVISED

Key facts

1 **Setting up equations**

A real-life problem is given in words. You need to write it in algebra. The first step is to decide on the variable and give it a letter, often x.
Then you form and solve an equation for x.
Always end with a statement in words.
Certain problems will lead to quadratic or cubic equations.
The process is just the same.

2 **Setting up simultaneous equations**

When there are two unknown variables you need to use two letters, for example x and y.
So you set up and solve simultaneous equations.

Your first line will often be 'Let $x = \ldots$' or equivalent.

The last line is not '$x = \ldots$' as this is the solution to your equation and not the solution to the problem.

In some problems involving quadratic (or cubic) equations, one (or more) of the answers may be impossible, so they all need to be checked.

Worked example

Setting up linear equations

1 Mr Smith is 5 years older than his wife. He was 26 when their twins were born.

The total sum of their ages this year is 91. Find the age of each member of the family.

Solution

Let Mr Smith's age be x.

Then Mrs Smith's age is $x - 5$

and the twins are $x - 26$.

The total age is 91.

So $x + (x - 5) + 2(x - 26) = 91$.

$\Rightarrow \quad 4x - 57 = 91$

$\Rightarrow \quad 4x = 148$

$\Rightarrow \quad x = 37$

Mr Smith is aged 37, Mrs Smith 32 and the twins 11.

State the variable to be used for the unknown.

Set up the equation.

Solve the equation for x.

Answer the question.

Worked examples

Setting up quadratic equations

2 A farmer has 20 metres of fencing and wants to enclose a rectangular area of $42\,\text{m}^2$ using the fencing for three sides and a long wall for the other side. What should be the dimensions of the rectangle?

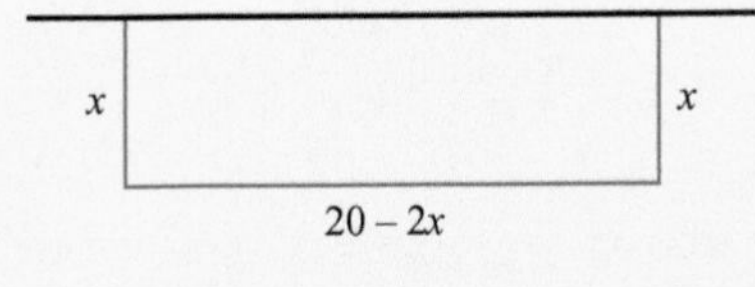

Solution

Let x m be the length of the two equal sides.

Then the length of the third side is $20 - 2x$ m.

Since the area is to be $42\,\text{m}^2$ this gives

$$x(20 - 2x) = 42$$

State the variable to be used for the unknown

Set up the equation

$\Rightarrow \quad 20x - x^2 = 42$

$\Rightarrow x^2 - 10x + 21 = 0$

$\Rightarrow (x-7)(x-3) = 0$

So $x = 3$ or 7.

If $x = 3$ then the sides are 3m, 3m, 14m

If $x = 7$ then the sides are 7m, 7m, 6m

Check $3 \times 14 = 42$ and $7 \times 6 = 42$.

Both give an area of $42\,m^2$ as required.

Solve the equation for x. In this case there are two valid answers.

3 A group of children share 120 sweets. If there were 4 fewer children, then each child would get 5 more sweets. How many children are there and how many sweets did they receive?

Solution

Let x be the number of children.

State the variable to be used for the unknown clearly.

Then the number of sweets each receives is $\frac{120}{x}$.

If there were $x - 4$ children the number of sweets would be $\frac{120}{x-4}$ which is 5 more.

Set up the equation.

$$\Rightarrow \quad \frac{120}{x} + 5 = \frac{120}{x-4}$$

$\Rightarrow 120(x-4) + 5x(x-4) = 120x$

Manipulate to give a standard quadratic equation.

$\Rightarrow \quad 5x^2 - 20x - 480 = 0$

$\Rightarrow \quad x^2 - 4x - 96 = 0$

$\Rightarrow \quad (x-12)(x+8) = 0$

$\Rightarrow \quad x = 12$ or -8

There are two answers.

Reject 2nd value as you cannot have a negative number of children.

Because of the context one answer is rejected.

So, there were 12 children and each had 10 sweets.

Check: 4 fewer children means 8 children who would then get 15 sweets, as required.

Worked example

Setting up simultaneous equations

4 In an interval of a concert John buys 3 coffees and 5 ice creams for his party. The total cost is £14.10. A few days later he buys 4 coffees and 1 ice cream and the cost is £10.30. Find the cost of a coffee and an ice cream.

Solution

Let the cost of a coffee be x pence and the price of an ice cream be y pence.

State the variables you want to use.

Then from the first occasion $3x + 5y = 1410$ ①

On the second occasion $4x + y = 1030$ ②

From the two pieces of information set up two equations in x and y.

$5 \times$ ②: $20x + 5y = 5150$

①: $3x + 5y = 1410$

Subtract $\quad 17x \quad = 3740$

$x = 220$

$\Rightarrow y = 150$

Solve simultaneously to find x and y.

So the cost of a coffee is £2.20 and the cost of an ice cream is £1.50.

Remember to answer the question!

Exam style question

TESTED

Simon and Gavin drive a distance of 140 km along a motorway, both at constant speed. Gavin's speed is v km per hour and Simon drives 5 km per hour faster than Gavin.

(i) Write down expressions in terms of v for the times, in hours, taken by Gavin and Simon.

Simon completes the journey in 15 minutes less than Gavin.

(ii) Explain why $\frac{140}{v} - \frac{140}{v+5} = \frac{1}{4}$ and show that this equation reduces to the equation $v^2 + 5v - 2800 = 0$.

(iii) Solve this equation to find v and hence the times taken by Simon and Gavin. Give your answers to the nearest minute.

Short answer on page 120

Full worked solution online

CHECKED ANSWER

3.3 Inequalities

REVISED

Key facts

1 **Linear inequalities**

Inequalities can be manipulated in the same way as equations except for one rule: if both sides are multiplied or divided by a negative number then the direction of the inequality changes.

Remembering the above rule, linear inequalities can be solved in the same way as equations.

2 **Solving quadratic inequalities graphically**

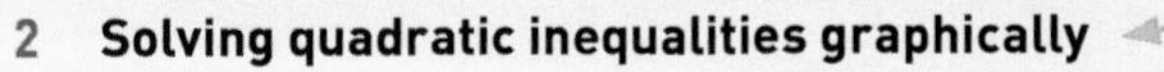

Always start by drawing a sketch graph of $y = f(x)$.

A quadratic inequality takes the form $f(x) < 0$ (with any of the four inequality signs), or can be manipulated into that form.
The graph represents the function $y = f(x) = ax^2 + bx + c$ where $a > 0$.

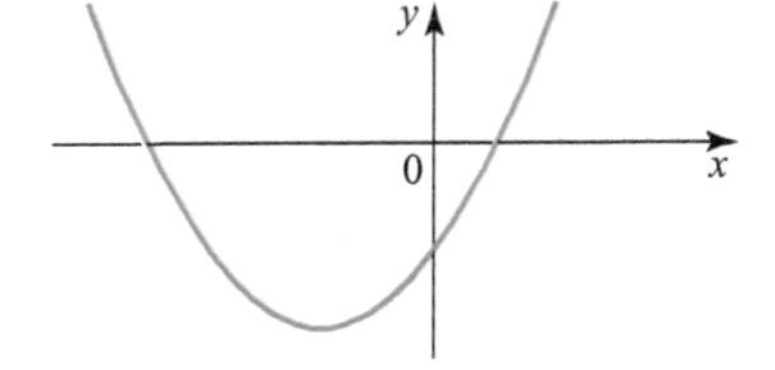

If $f(x) < 0$ then the range of values of x is the range where the graph is below the axis.

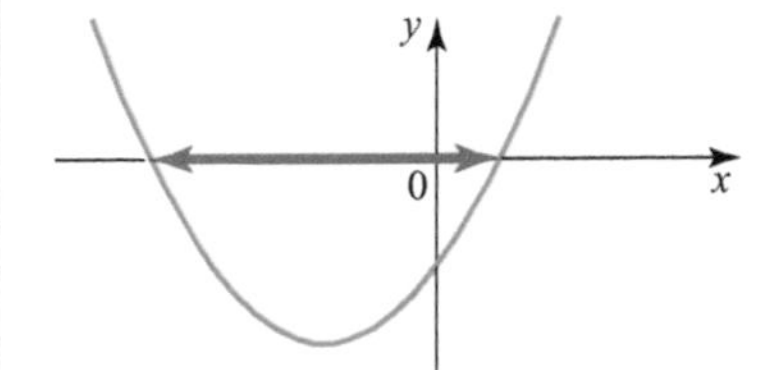

If $f(x) > 0$ then the range of values of x is the range where the graph is above the axis.

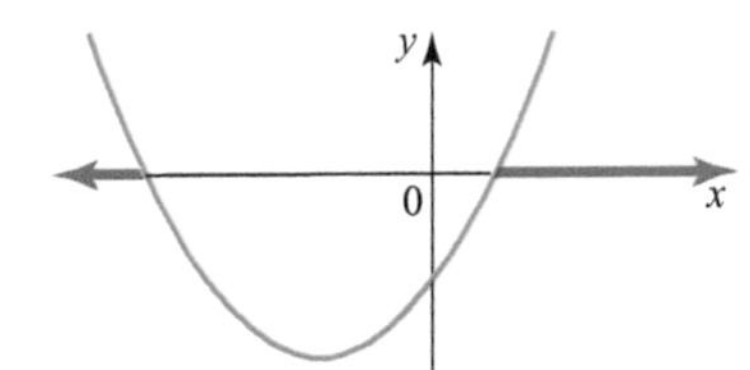

3 **Solving quadratic inequalities algebraically**

Start by solving the quadratic equation $f(x) = 0$.

That will give you roots p and q with $p < q$ (say). Then decide which regions $x < p$, $p < x < q$ and $x > q$ satisfy the inequality.

When $f(x)$ is factorised you have two expressions whose product satisfies the inequality. You can then use the facts:

+ve × +ve = +ve, −ve × −ve = +ve,

+ve × −ve = −ve −ve × +ve = −ve

to decide which regions satisfy the inequality.

Worked examples

Linear inequalities

1 Solve the inequality $2 - x < 3(2x - 1)$.

Solution

$2 - x < 3(2x - 1)$

$2 - x < 6x - 3$ ← Expand the bracket.

$5 < 7x$ ← Collect terms in x on one side and the numbers on the other. It does not matter which side.

$\frac{5}{7} < x$ ← Divide by 7 to get x.

$x > \frac{5}{7}$ ← Write x on the left side of the inequality. Logically $x > \frac{5}{7}$ is the same as $\frac{5}{7} < x$.

Sometimes, as in the next example, you will get a double inequality. Apply the same rules throughout.

2 Find the range of values of x for which $3 < 2x - 7 < 9$.

Solution

$3 < 2x - 7 < 9$

$\Rightarrow \quad 10 < 2x < 16$ ← Add 7 to all three parts.

$\Rightarrow \quad 5 < x < 8$ ← Divide through by 2.

Worked example

Solving quadratic inequalities graphically

3 Solve the inequality $x^2 + 2x - 8 > 0$ graphically.

Solution

The graph of $y = x^2 + 2x - 8$ is shown below.

It crosses the x-axis at -4 and $+2$.

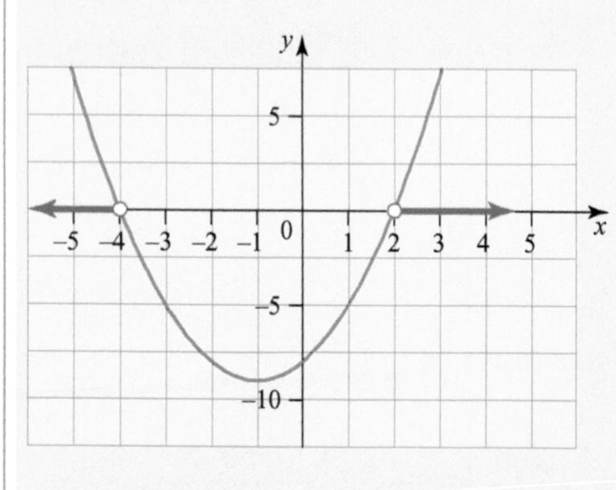

The graph is above the x-axis (i.e. $y > 0$) for values of x greater than 2 and less than –4.

Note that if the inequality is $f(x) \geqslant 0$ then the end points are included but if the inequality is $f(x) > 0$ then they are not.

If they are included the standard notation is for a filled circle at the end of the line •.

but if they are not then the circle is not shaded (is 'open') ∘.

In this example the end points are not included.

So the solution is $x < -4$ or $x > 2$.

Common mistake: It would not be correct to write $-4 > x > 2$.

Worked example

Solving quadratic inequalities algebraically

4 Solve the inequality $x^2 - 4x - 32 < 0$.

Solution

Start by solving the equation

$$x^2 - 4x - 32 = 0$$

$\Rightarrow \quad (x+4)(x-8) = 0$

So $x = -4$ or $x = 8$

Factorise the quadratic equation.

Now consider the three regions

$x < -4$, $-4 < x < 8$, $x > 8$

For $x < -4$, $(x + 4)(x - 8)$ is $- \times - = +$.

The inequality is not satisfied.

For $-4 < x < 8$, $(x + 4)(x - 8)$ is $+ \times - = -$.

The inequality is satisfied.

For $x > 8$, $(x + 4)(x - 8)$ is $+ \times + = +$.

The inequality is not satisfied.

So the solution is $-4 < x < 8$.

Exam-style question

TESTED

(i) Solve algebraically the inequality $x^2 + 2x - 15 \geqslant 0$.

(ii) Illustrate your answer graphically.

Short answers on page 120

Full worked solutions online

CHECKED ANSWERS

Chapter 4 Recurrence relationships

About this topic

A sequence is an ordered (possibly infinite) set of objects or numbers. In this course you will be concerned with sequences of numbers.

Each number in a sequence is called a term and is identified by its position within the sequence.

A recurrence relationship is a rule that connects two or more terms.

A finite sequence (an ordered set of numbers) can simply be listed, but an infinite sequence requires a rule.

You often meet sequences in everyday life, for example measurements made at regular intervals of time or distance.

Before you start, remember ...

- algebraic manipulation from GCSE
- the use of suffices to distinguish terms, e.g. x_1, x_2.

4.1 First order recurrence relationships

REVISED

Key facts

1 **Position to term sequences**

Each term is defined by a rule that connects it to the value of *n*, its position in the sequence. So each term can be worked out directly.

For example: for $u_n = n^2 + 3$, $u_6 = 6^2 + 3 = 39$.

2 **Term to term sequences**

Each term is defined by a rule that connects it with one of the previous terms. In this case the first term is required as well as the rule.

For example: $u_{n+1} = u_n + 3$ with $u_1 = 1$.

The rule for a term to term sequence is called a recurrence relationship

3 **The order of a recurrence relationship**

A first order recurrence relationship connects a term with the previous term.

A second order recurrence relationship connects a term with the previous two terms.

And so on.

Worked example

Position to term sequence

1 List the first five terms of the sequence given by $x_n = 3n^2 - 1$.

Each term is calculated from n, the position of the term in the sequence.

Solution

$x_1 = 3 \times 1^2 - 1 = 2$

So x_1 is found by substituting $n = 1$.

$x_2 = 3 \times 2^2 - 1 = 11$

x_2 by substituting $n = 2$, and so on.

$x_3 = 3 \times 3^2 - 1 = 26$

$x_4 = 3 \times 4^2 - 1 = 47$

$x_5 = 3 \times 5^2 - 1 = 74$

Worked example

Term to term sequence

2 List the first five terms of the sequence given by $x_1 = 1$ and $x_n = 3x_{n-1} - 1$.

Solution

$x_2 = 3 \times x_1 - 1 = 3 \times 1 - 1 = 2$

You are given x_1. Now use the rule to work out x_2, etc.

$x_3 = 3 \times x_2 - 1 = 3 \times 2 - 1 = 5$

$x_4 = 3 \times x_3 - 1 = 3 \times 5 - 1 = 14$

$x_5 = 3 \times x_4 - 1 = 3 \times 14 - 1 = 41$

Worked example

Using recurrence relationships in modelling

3 In an enclosed space there are initially 10 bacteria. A scientist believes that the number of bacteria can be modelled using the assumption that the number of bacteria will double every hour.

(i) According to the model, how long is it before the number of bacteria will exceed 1000?

(ii) Why, in the long term, might the model be inappropriate?

Solution

(i) Let x_r be the number of bacteria at the beginning of the rth hour.

So $x_1 = 10$ and $x_{r+1} = 2x_r$

The recurrence relationship is that in each hour the number doubles.

$x_1 = 10 \Rightarrow x_2 = 20,\ x_3 = 40,\ x_4 = 80,\ x_5 = 160,$

Work out each term.

$x_6 = 320,\ x_7 = 640,\ x_8 = 1280$

So by the beginning of the 8th hour the number of bacteria will have exceeded 1000; after 7 hours the number is greater than 1000.

(ii) The model predicts that the number of bacteria will continually double every hour, which is not sustainable.

Exam-style question

TESTED

Ahmed takes out a loan of £2000 on 1 June 2021. This is denoted by £L_1. The amount of his loan at the end of 1 June 2022 is £L_2; at the end of 1 June 2023 is £L_3 and so on.

The amount of his loan is increased by 10% interest at the start of 1 June each year.

Ahmed agrees to pay back a fixed amount at midday on 2 June each year, starting on 2 June 2022.

(i) Write down a recurrence relationship giving L_{n+1} in terms of L_n for $n \geqslant 1$.

(ii) What is the minimum amount that Ahmed must pay each year so that the loan does not increase over time?

(iii) If he makes no repayments, how much does he owe after 5 years?

Short answers on page 120

Full worked solutions online

CHECKED ANSWERS

4.2 Second order recurrence relationships

REVISED

Key fact

1 **A recurrence relationship is called second order if the relationship involves the previous two terms. In this case, two terms need to be given.**
For example: $u_{n+1} = 2u_n + u_{n-1}$ with $u_1 = 1$, $u_2 = 2$.

Worked example

Second order relationships

1 A sequence of terms u_n is defined by $u_{n+1} = 2u_n - u_{n-1}$.

Find u_n for $n = 3, 4, 5$ and 6 when

(i) $u_1 = u_2 = 1$,

(ii) $u_1 = 1$, $u_2 = 2$.

(iii) In each case make a conjecture about the value of u_n.

Solution

(i) $u_3 = 2u_2 - u_1 = 2\times1 - 1 = 1$

$u_4 = 2u_3 - u_2 = 2\times1 - 1 = 1$

Similarly $u_5 = u_6 = 1$

(ii) $u_3 = 2u_2 - u_1 = 2\times2 - 1 = 3$

$u_4 = 2u_3 - u_2 = 2\times3 - 2 = 4$

$u_5 = 2u_4 - u_3 = 2\times4 - 3 = 5$

$u_6 = 2u_5 - u_4 = 2\times5 - 4 = 6$

(iii) In (i) $u_n = 1$, in (ii) $u_n = n$.

Exam-style question

TESTED

A sequence of terms is defined by $u_{n+1} = 4u_n - 3u_{n-1}$ with $u_1 = 1$ and $u_2 = 2$.

(i) Find the first five terms of the sequence.

(ii) Amit suggests that the terms of the sequence can be given by $u_n = \frac{1}{2}\left(1+3^{n-1}\right)$. Show that this rule is correct for the first 5 terms.

Short answers on page 120

Full worked solutions online

CHECKED ANSWERS

Review questions (Chapters 1–4)

1 Simplify $(x-3)(x+2)+(x+3)(x+2)$. **[2]**

2 Express $(\sqrt{3}+3\sqrt{5})(2\sqrt{3}-\sqrt{5})$ in the form $a+b\sqrt{c}$ where a, b and c are integers to be determined. **[2]**

3 Simplify $\frac{1}{2+\sqrt{2}}+\frac{1}{2-\sqrt{2}}$. **[2]**

4 Express $\frac{3\sqrt{3}-\sqrt{2}}{2\sqrt{3}+\sqrt{2}}$ in the form $a+b\sqrt{n}$ where a and n are integers and b is a rational number. **[4]**

5 Find the quotient when x^3-x^2+3x+4 is divided by $(x+2)$. **[2]**

6 Solve the equation $\frac{2(x-1)}{3}-2=\frac{3x+1}{4}$. **[3]**

7 Solve the equation $x^2-4x-7=0$.
Give your answers correct to three significant figures. **[3]**

8 Solve the equation $\frac{3}{x-1}+\frac{2}{x-2}=1$ giving the roots exactly. **[4]**

9 You are given $f(x)=x^3-8x^2+5x+14$.

(i) Show that $(x-2)$ is a factor of $f(x)$. **[2]**

(ii) Solve the equation $f(x)=0$. **[3]**

10 You are given that the cubic equation $x^3+ax^2+bx-22=0$ has three distinct, positive roots.
By forming two equations in a and b, find the values of a and b. **[5]**

11 Solve simultaneously the equations $y=x+3$ and $y=x^2-2x+5$. **[4]**

12 Four years ago Jean was 6 times as old as her daughter Carla. This year she is 4 times as old.
Taking Carla's age now as x years, form an equation in x and solve to find Carla's age now. **[4]**

13 I regularly travel a journey of 200 kilometres. When I travel by day, I average v kilometres per hour. When I travel at night, there is less traffic so I can average 20 kilometres per hour faster. This means that I can complete the journey in 50 minutes less time.

(i) Write down expressions for the journey times during the day and at night. **[2]**

(ii) Hence form an equation in v and show that it simplifies to $v^2+20v-4800=0$. **[4]**

(iii) Hence find the times it takes me to complete the journey during the day and at night. **[5]**

14 Solve the following inequalities.

(i) $3-x>5(x+1)$ **[3]**

(ii) $x^2+5x<6$ **[3]**

15 Find the set of integers that satisfy the inequality $-8<3x-1<13$. **[3]**

16 £2000 is deposited into a savings account on 1 January one year. On 1 January the next year, the amount has become £2100.

(i) What was the rate of interest in that year? **[2]**

(ii) On the assumption that the rate of interest remains constant, find the amount in the account after 3 years. **[2]**

17 A sequence of positive integers, $u_1, u_2, u_3, \ldots$ is such that $u_{n+2}=5u_{n+1}-6u_n$. with $u_1=1$ and $u_2=5$.
Philip proposes that the formula for this sequence is $u_n=A\times2^n+B\times3^n$.

(i) Find the values of A and B. **[3]**

(ii) Calculate the values of u_3 and u_4. **[2]**

Short answers on pages 120–21

Full worked solutions online

CHECKED ANSWERS

Section 2 Coordinate geometry in two dimensions

Target your revision (Chapters 5–7)

Try answering each question below. If you get stuck, follow the page reference underneath to revise that topic.

1 **Points and lines**
Find the equation of the line AB where A and B have coordinates (–2, 3) and (4, –2).
(see page 28)

2 **The gradient of a line**
Find the gradient of the line $2x + 5y = 7$.
(see page 28)

3 **Parallel and perpendicular lines**
Find the equations of the lines through the point (1, 2) which are parallel to and perpendicular to the line $3x + 4y = 6$.
(see page 28)

4 **The circle**
Find the centre and radius of the circle whose equation is

$x^2 + y^2 - 4x + 6y = 3$
(see page 30)

5 **The equation of a line and its graphical representation**
Draw the line $2x + 5y = 20$.
(see page 33)

6 **Plotting or sketching a quadratic function**
Sketch the graph of $y = x^2 + 4x - 7$.
(see page 33)

7 **Plotting or sketching polynomial functions**
Plot the graph of $y = x^3 - 2x^2 - x + 1$ for values of x from $-2 \leqslant x \leqslant 3$.
(see page 34)

8 **Trigonometrical functions**
Plot the graph of $y = \cos x + \sin x$ for $0° \leqslant x \leqslant 360°$.
(see page 36)

9 **Exponential functions**
Plot the graph of $y = \frac{3^x}{2} - 1$ for $-1 \leqslant x \leqslant 3$.
(see page 36)

10 **Linear inequalities in two variables**
(i) The variables x and y are non-negative numbers. Express this as two inequalities.
(ii) Additionally, x and y are to obey the following three inequalities.

$2x + 3y \leqslant 24$

$x \geqslant 2$

$5x + 2y \leqslant 30$

Illustrate these inequalities on one graph. Shade the region of the graph that does not satisfy the conditions.
(iii) Find the maximum value of $2x + y$ in the region if x and y are to be integers.
(see page 39)

Short answers on page 121
Full worked solutions online

CHECKED ANSWERS ☐

Chapter 5 Points, lines and circles

About this topic

Coordinate geometry provides the link between geometry and algebra. Being able to use algebra makes it much easier to solve many geometric problems. In this chapter, you meet the coordinate geometry of straight lines and circles.

Before you start, remember ...

- how to use coordinates
- Pythagoras' theorem
- how to solve linear equations.

5.1 Points and lines

REVISED

Key facts

1 **The gradient of a line**
The gradient of a line joining the points (x_1, y_1) and (x_2, y_2) is

$$m = \frac{y_2 - y_1}{x_2 - x_1}.$$

This is $\frac{\text{change in } y \text{ values}}{\text{change in } x \text{ values}}$.

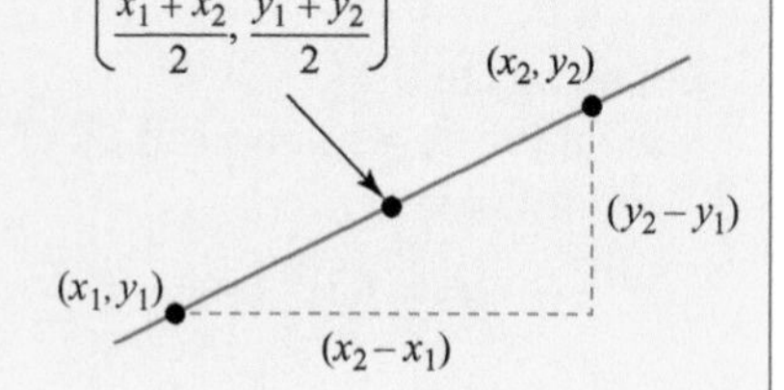

2 **The distance between two points**
The distance between two points

(x_1, y_1) and (x_2, y_2) is $\sqrt{(x_2 - x_1)^2 + (y_2 - y_1)^2}$.

This is Pythagoras' theorem.

3 **The midpoint of a line joining two points**
The midpoint of the line between (x_1, y_1) and (x_2, y_2) is

$$\left(\frac{x_1 + x_2}{2}, \frac{y_1 + y_2}{2}\right).$$

The coordinates of the midpoint of a line are the means of the end points.

4 **Parallel lines**
Parallel lines have the same gradient.

5 **Perpendicular lines**
Lines with gradients m_1 and m_2 are perpendicular if $m_1 m_2 = -1$.

Note that unless you scale the axes equally perpendicular lines may not look as if they are at right angles to each other.

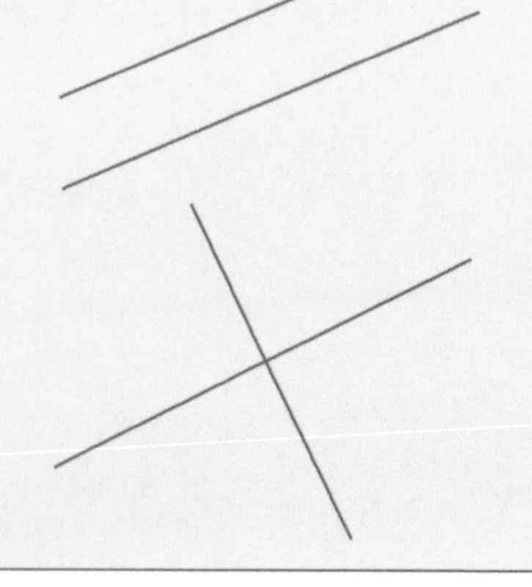

Worked example

Gradients of lines

1 Find the gradient of the line joining the two points (1, 2) and (7, 9).

Solution

$$m = \frac{9-2}{7-1}$$

The gradient is $\frac{7}{6}$.

Using $m = \frac{y_2 - y_1}{x_2 - x_1}$.

Worked example

The distance between two points

2 Find the distance between the points A(5, −3) and B(−2, 6).

Solution

By Pythagoras, distance $AB = \sqrt{(5 - -2)^2 + (-3 - 6)^2}$

$= \sqrt{(7)^2 + (-9)^2} = \sqrt{49 + 81} = \sqrt{130}$.

Using:
$\text{distance} = \sqrt{(x_2 - x_1)^2 + (y_2 - y_1)^2}$
Take care with the signs.

Worked example

The midpoint of a line

3 Find the midpoint of the line PQ where P is (−3, 7) and Q is (4, −1).

Solution

The midpoint is

$$\left(\frac{-3+4}{2}, \frac{7-1}{2}\right) = \left(\frac{1}{2}, 3\right).$$

Using:
$\text{midpoint} = \left(\frac{x_1 + x_2}{2}, \frac{y_1 + y_2}{2}\right)$.

Worked example

Parallel and perpendicular lines

4 The coordinates of L, M and N are (2, 1), (6, 3) and (1, 3), respectively.

Show that the lines LM and LN are perpendicular.

Solution

Gradient of the line $LM = m_1 = \frac{3-1}{6-2} = \frac{2}{4} = \frac{1}{2}$

Gradient of the line $LN = m_2 = \frac{3-1}{1-2} = \frac{2}{-1} = -2$

$$\Rightarrow m_1 m_2 = \frac{1}{2} \times -2 = -1$$

So LM and LN are perpendicular.

Find the gradient of each line.

Checking whether $m_1 m_2 = -1$.

Exam-style question

TESTED

Points A, B, C and D have coordinates (1, 0), (2, 2), (4, 1) and (2, −3) respectively.

(i) Show that the quadrilateral ABCD is a trapezium.
(ii) E is on the line CD with coordinates (3, −1).
Show that the quadrilateral ABCE is a parallelogram.
(iii) Show additionally that the quadrilateral ABCE is a rhombus.
(iv) Prove that ABCE is a square.

Short answers on page 122
Full worked solutions online

CHECKED ANSWERS

5.2 The circle

REVISED

Key facts

1 Definition
A circle is the locus of all points that are a fixed distance (the radius) from a fixed point (the centre).

A circle is a 2-dimensional shape. In 3 dimensions this would be the definition of a sphere.

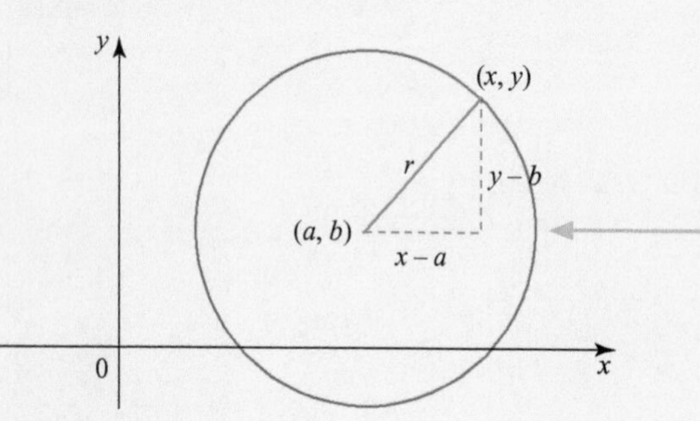

Caution: If you try plotting this curve on your graphic calculator
(i) it will not look like a circle if the axes have different scales
(ii) your calculator may only show half the circle.

2 The equation of a circle with given centre and radius
The equation of a circle with centre (a, b) and radius is r is

$$(x-a)^2+(y-b)^2=r^2$$

This equation is a statement of Pythagoras theorem.

3 Equation of a circle with centre at the origin
For a circle with centre the origin, (a, b) is $(0, 0)$ and so the equation is

$$x^2+y^2=r^2$$

4 The general equation of a circle
An alternative general form of the equation of a circle is

$$x^2+y^2+2fx+2gy+c=0$$

The centre is at $(-f, -g)$ and the radius is r where $r^2=f^2+g^2-c$.

Note that
- the coefficients of the x^2 and y^2 terms are equal,
- there is no xy term.

Worked example

The equation of a circle

1 Find the equation of the circle with centre (1, 2) and radius 4.

Solution

Always start by drawing a diagram.

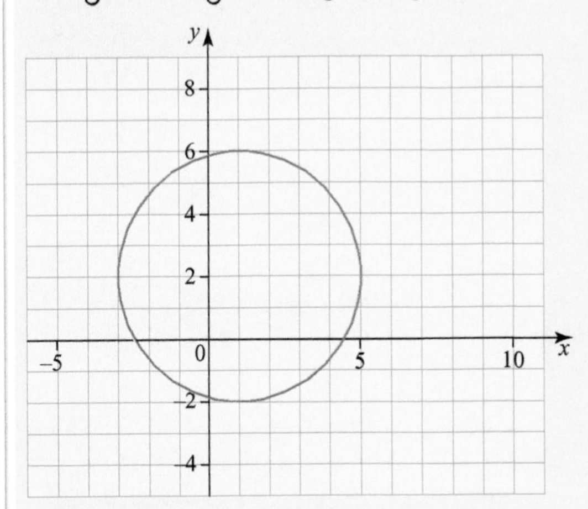

Applying the formula $(x-a)^2+(y-b)^2=r^2$ gives

$(x-1)^2+(y-2)^2=4^2$

If you are asked for the equation of a circle in an examination, then either form will be acceptable unless one specific form is demanded.

When expanded this is

$x^2-2x+1+y^2-4y+4=1$

and this can be simplified to

$x^2+y^2-2x-4y-11=0$

In the previous example you were given a circle and asked to find its equation. The next example is the other way round – you are given the equation of a circle and are asked to find its centre and radius.

Worked example

Finding the centre and radius of a circle

2 Use the method of completing the square to find the centre and radius of the circle $x^2+y^2-4x+6y+4=0$.

Solution

$$x^2+y^2-4x+6y+4=0$$

$$\Rightarrow \quad (x^2-4x)+(y^2+6y)+4=0$$

Complete the square for both x and y.

$$\Rightarrow \quad (x^2-4x+4)+(y^2+6y+9)+4-4-9=0$$

$$\Rightarrow \quad (x-2)^2+(y+3)^2=9$$

So the centre of the circle is (2, −3) and the radius is 3.

Worked example

Points inside or outside the circle

3 Using the general form of the equation of a circle, determine whether the point (4, –6) lies inside or outside the circle with equation $x^2 + y^2 - 6x + 8y + 19 = 0$.

Solution

Comparison with the general equation for a circle

$x^2 + y^2 + 2fx + 2gx + c = 0$

shows that the centre of the circle is at (3, –4) and the radius, r, is given by

$r^2 = f^2 + g^2 - c = 3^2 + 4^2 - 19 = 6$

To find distance, d, from the centre of the circle to the point, use Pythagoras.

$d^2 = (4-3)^2 + (-6--4)^2 = 1 + 4 = 5$

$d^2 < r^2$

so the point is inside the circle.

Find the centre of the circle and the radius.

Note that finding r^2 is sufficient in this example.

The distance is less than the radius so it lies inside the circle.

Exam-style question

TESTED

A circle has centre (0, 1) and radius 3.

(i) Find the equation of the circle.
(ii) The points A and B are where the circle cuts the x-axis. Find the length AB.
(iii) Determine whether the point (2, 3.5) lies inside, outside or on the circle.

Short answer on page 122

Full worked solution online

CHECKED ANSWERS

Chapter 6 Graphs

About this topic

A function f(x) can be illustrated as a curve by drawing the graph of $y = f(x)$. This geometrical representation is really important. It allows you to visualise the function and highlights its main features.

Before you start, remember ...

- the difference between a plot and a sketch
- polynomials (from Chapter 2)
- coordinate geometry (from Chapter 5).

6.1 The equation of a line and its graphical representation

REVISED

Key facts

1. **The equation of a line**
   There are a number of different forms of the equation of a line.
   The most basic form is $y = mx + c$.
   The gradient of this line is m and the intercept on the y-axis is $(0, c)$.
2. **The equation of a line through a given point with given gradient**
   The equation of a line with gradient m through the point (x_1, y_1) is
   $y - y_1 = m(x - x_1)$
3. **The equation of a line through 2 points**
   The equation of the line through the two points (x_1, y_1) and (x_2, y_2) is
   $\frac{y - y_1}{x - x_1} = \frac{y_2 - y_1}{x_2 - x_1}$
4. **Drawing a straight line**
   You need to know two points to draw a straight line. A third point can be used as a check.

You find this by substituting $x = 0$ into the equation.

This is because a general point is (x, y) and if it lies on the line then the gradient from this point to the given point (x_1, y_1) is m.
So $m = \frac{y - y_1}{x - x_1}$.

This is because both sides of the equation are equal to m.

Worked examples

Equation of a line

1 Find the equation of the line with gradient 2 and which passes through the point (0, −2).

Solution

The most convenient form in this case is $y = mx + c$ because c is the intercept on the y-axis.

So $m = 2$ and $c = -2 \Rightarrow y = 2x - 2$

Warning: You should multiply out and simplify any equations of lines so that there are three terms. Appropriate forms are $y = mx + c$ and $ax + by = c$.

2 Find the equation of the line with gradient 3 and which passes through the point (1, 4).

Solution

The most convenient form in this case is $y - y_1 = m(x - x_1)$

(x_1, y_1) is (1, 4) and $m = 3$

This gives $y - 4 = 3(x - 1)$

$\Rightarrow \quad y - 4 = 3x - 3$

$\Rightarrow \quad y = 3x + 1$

3 Find the equation of the line through (2, 3) and (7, 10).

Solution

The most convenient form in this case is $\dfrac{y - y_1}{y_2 - y_1} = \dfrac{x - x_1}{x_2 - x_1}$.

(x_1, y_1) is (2, 3) and (x_2, y_2) is (7, 10).

This gives $\dfrac{y - 3}{10 - 3} = \dfrac{x - 2}{7 - 2}$

$\Rightarrow \quad \dfrac{y - 3}{7} = \dfrac{x - 2}{5}$

$\Rightarrow \quad 5y - 15 = 7x - 14$

$\Rightarrow \quad 5y = 7x + 1$

Caution: Notice that the penultimate lines of examples 2 and 3 would not be acceptable as final answers.

Exam-style question

TESTED

Plot the line $4x - 3y = 12$.

Short answer on page 122

Full worked solution online

CHECKED ANSWERS

6.2 Plotting or sketching polynomial functions

REVISED

Key facts

1 **Polynomial functions**
Linear, quadratic and cubic functions are all examples of polynomials.

2 **Plotting curves**
To **plot** a curve
- make out a table of values
- mark the points on your graph
- join them with a smooth curve.

To **sketch** a curve
- use your knowledge of the functions to draw its approximate shape
- mark on important points like where it crosses the axes and any turning points it has.

To **draw** a curve
- you are expected to make decisions about how accurate it should be; this will depend on the context
- usually the instruction 'draw' is taken to suggest an accuracy somewhere between plot and sketch.

In this course these are the only polynomials you will meet.

A turning point is an example of a stationary point.

Worked example

Plotting a quadratic function

1 Plot the curve $y = x^2 - 5x + 7$ in the range $-1 \leqslant x \leqslant 5$.

Solution

Points on the curve:

x	−1	0	1	2	3	4	5
y	13	7	3	1	1	3	7

Calculate the values for y given the value for x on your calculator.

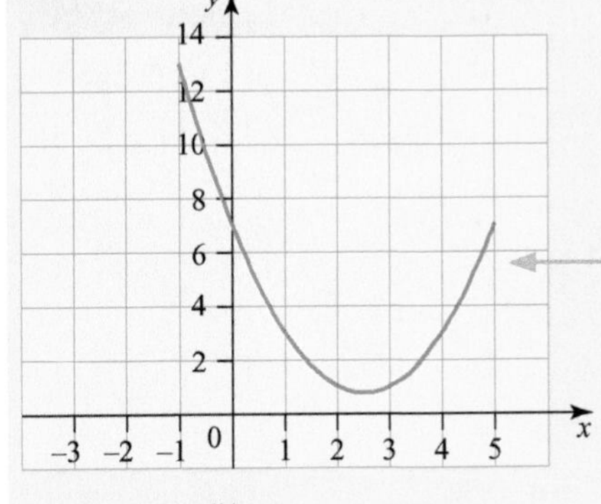

Note that a quadratic curve is symmetric in the line through the turning point parallel to the y-axis.

Note that a quadratic curve can be this way up or 'upside down' as in the graph below.

The curve will be this way up if the coefficient of the term in x^2 is negative.

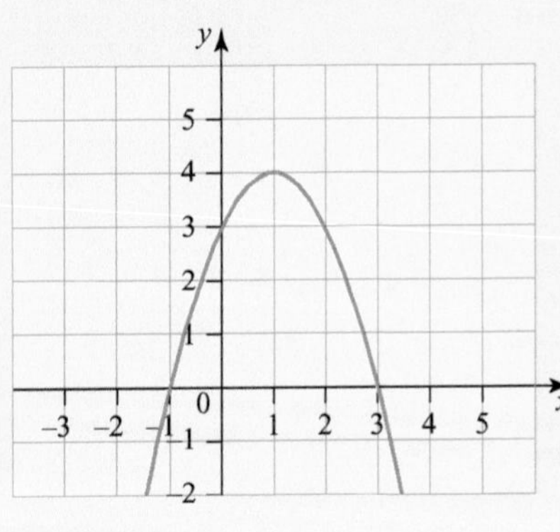

Worked example

Plotting a cubic curve

2 Plot the curve $y = x^3 - 3x^2 + x + 3$ for $-1 \leqslant x \leqslant 3$.

Solution

Points on the curve:

x	−1	0	1	2	3
y	−2	3	2	1	6

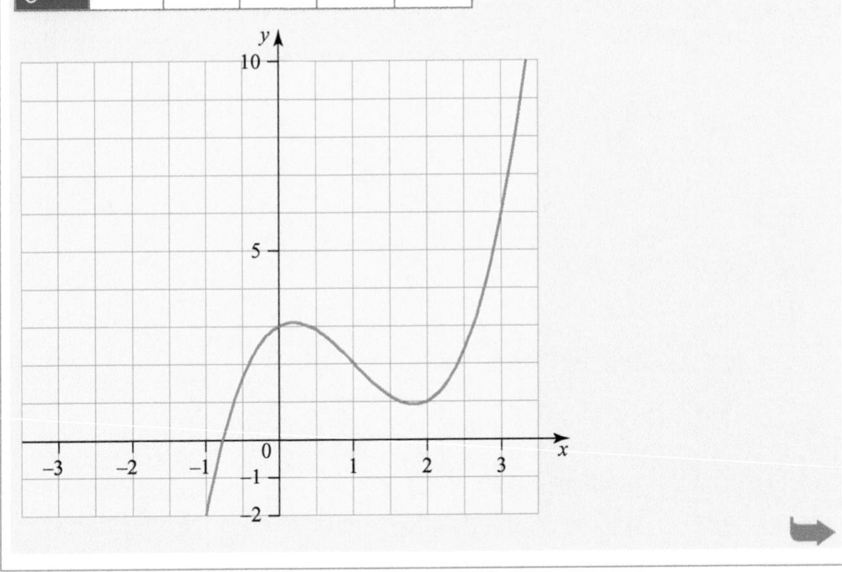

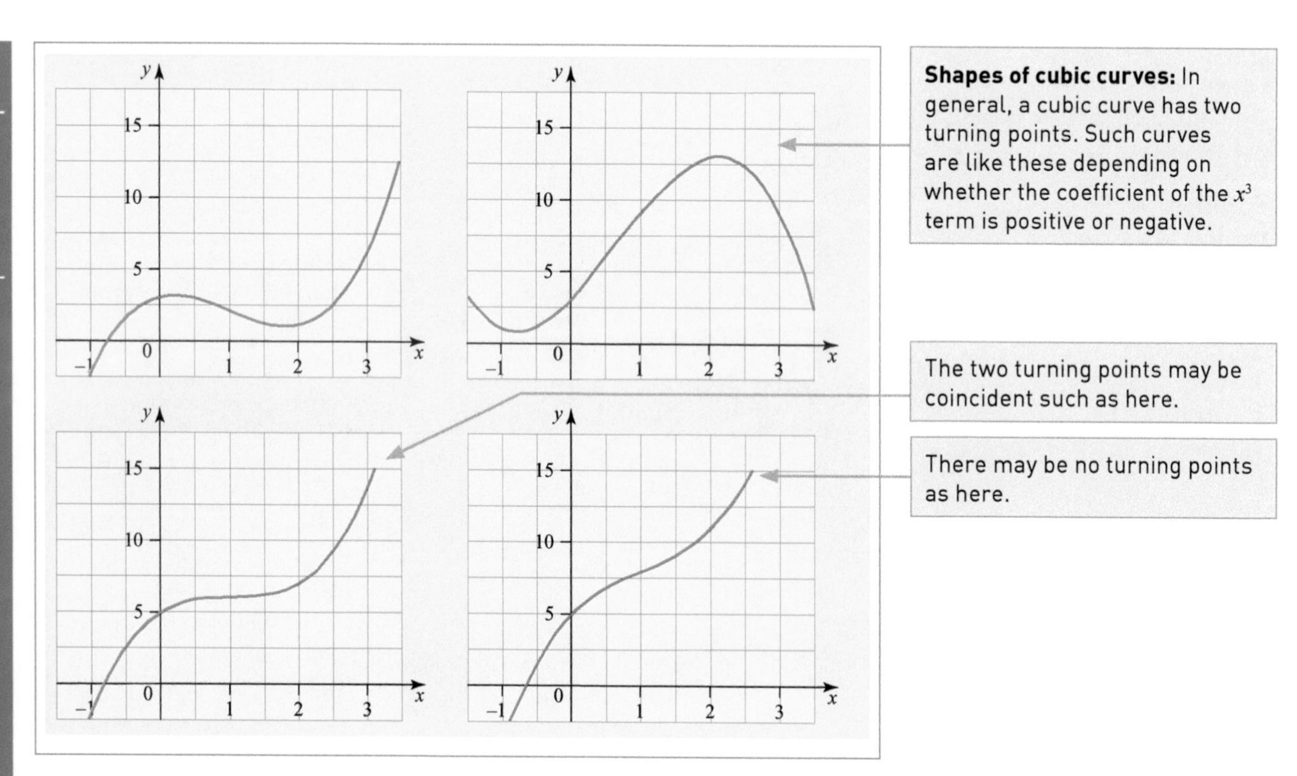

Exam-style question

TESTED

(i) Sketch on the same grid the curves with the following equations.

$y = x^3 - 6x^2 + 11x - 3$

$y = -x^3 + 6x^2 - 11x + 9$

(ii) Find algebraically the x-coordinates of the points of intersection of the curves.

Short answer on page 122

Full worked solution online

CHECKED ANSWERS

6.3 Trigonometric and exponential functions

REVISED

Key facts

1 Trigonometrical functions

You have seen the three ratios $\sin\theta$, $\cos\theta$ and $\tan\theta$ used in the context of right-angled triangles, and so for angles less than 90°.
However, your calculator will give you values for angles greater than 90°.
So you can plot the graphs of trigonometric functions for a greater range of values.
You need to know the graphs of the ratios from 0° to 360°.

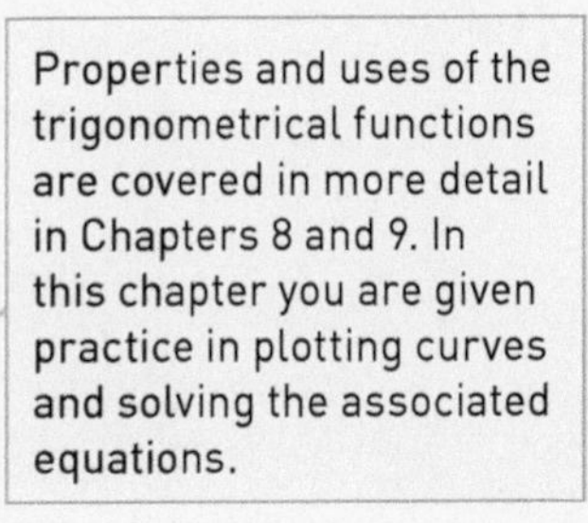
Properties and uses of the trigonometrical functions are covered in more detail in Chapters 8 and 9. In this chapter you are given practice in plotting curves and solving the associated equations.

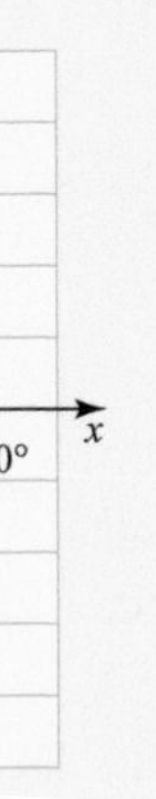

$y = \sin\theta$

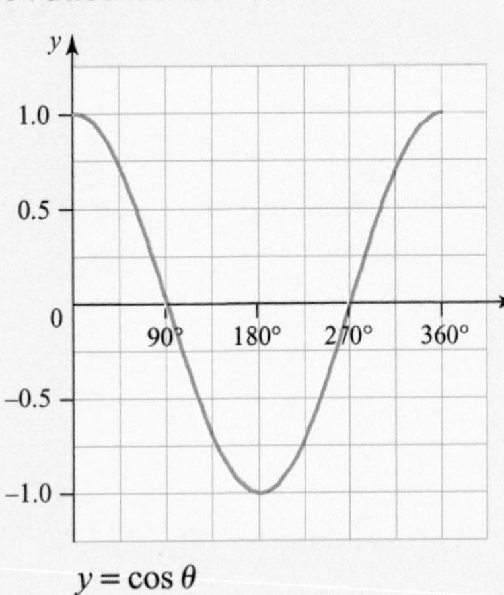

$y = \cos\theta$

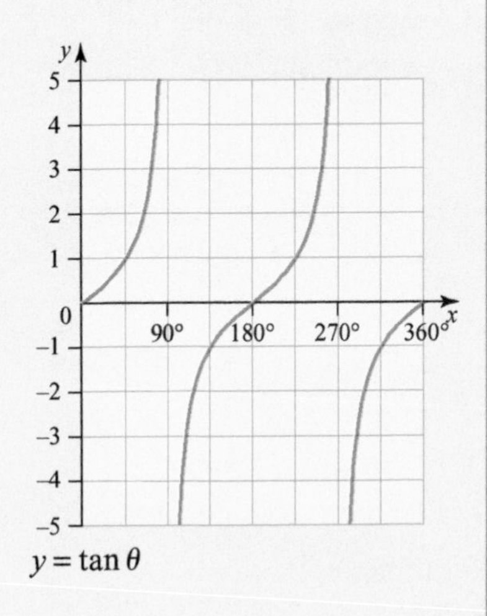

$y = \tan\theta$

2 Exponential functions

An exponential function is one where the variable is the power, e.g. $y = 3^x$.
Properties of the exponential functions of the form $y = ka^x$ where $a > 1$ and k is a positive number are

- y is always positive
- the curve passes through the point $(0, k)$
- the gradient is always positive and increases for increasing x
- the curve approaches the negative x-axis.

Properties and uses of the exponential functions are covered in more detail in Chapter 12.

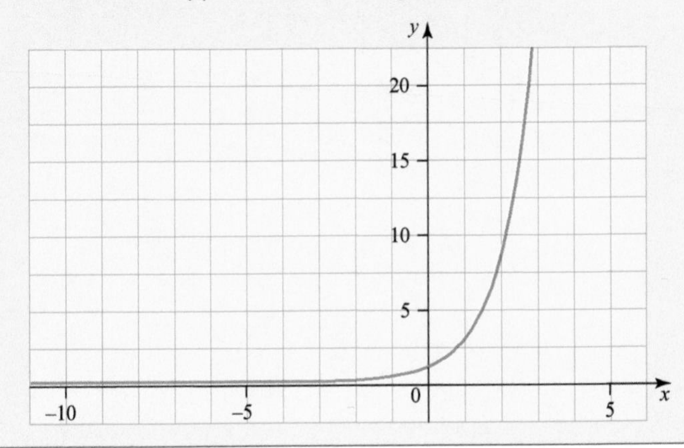

An equivalent exponential function is of the form $y = a^{-x}$ where $a > 1$.

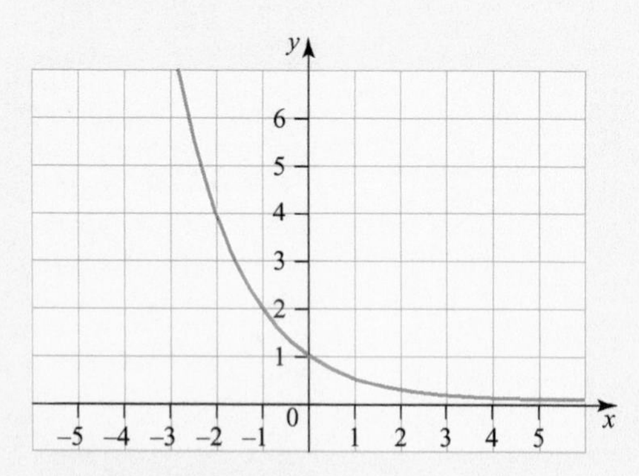

In this case

- y is always positive
- the curve passes through the point (0, 1)
- the gradient is always negative and increasing for increasing x.
- the curve approaches the positive x-axis.

Worked example

Trigonometric graphs

1 Plot the curves $y = \sin\theta$ and $y = \cos\theta$ on the same graph for $0° \leq \theta \leq 360°$.

Solution

θ	0	30	60	90	120	150	180	210	240	270	300	330	360
$\sin\theta$	0	0.5	0.87	1	0.87	0.5	0	−0.5	−0.87	−1	−0.87	−0.5	0
$\cos\theta$	1	0.87	0.5	0	−0.5	−0.87	−1	−0.87	−0.5	0	0.5	0.87	1

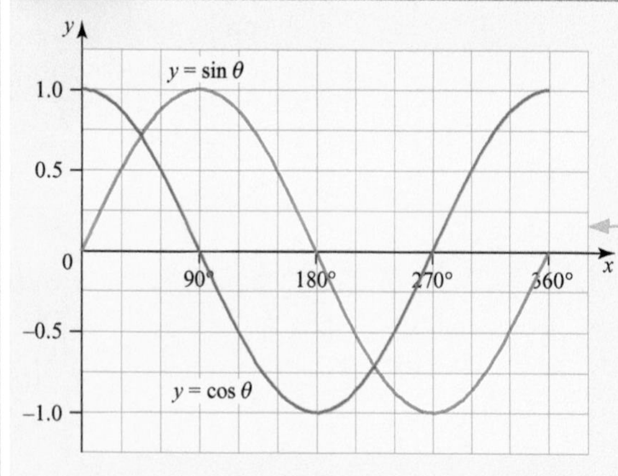

Notice that the two curves have the same shape but that $y = \cos\theta$ is translation of $y = \sin\theta$ through $-90°$ (i.e. $90°$ to the left).

Worked example

Exponential graphs

2 Plot the graph of $y = 3^{-x}$ in the range $x = -2$ to $x = 2$.

> Notice that $3^{-x} = \left(\frac{1}{3}\right)^x$ and so an alternative way of writing the equation of this curve is $y = \left(\frac{1}{3}\right)^x$.

Solution

x	−2	−1	0	1	2
3^{-x}	9	3	1	0.33	0.11

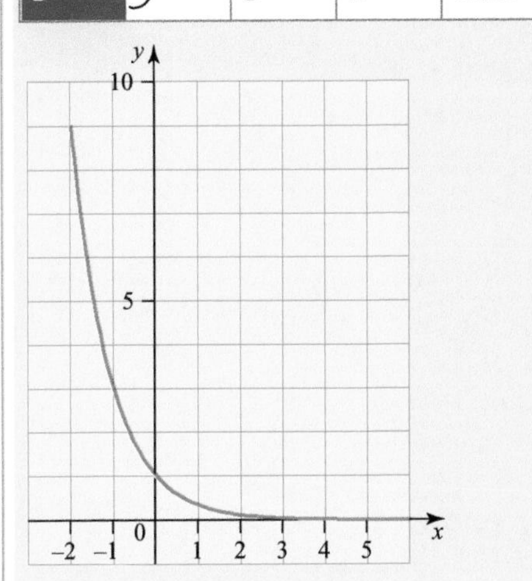

Exam-style question

TESTED

(i) Plot on the same grid the curves $y = 2^x + 1.5$ and $y = 3^x$.
(ii) From your graph estimate the root of the equation $2^x + 1.5 = 3^x$.
(iii) Substitute the value of x you obtained in part (ii) in each of the equations $y = 2^x + 1.5$ and $y = 3^x$. Hence check your answer to part (ii).

Short answer on page 122
Full worked solution online

CHECKED ANSWERS

Chapter 7 Linear inequalities in two variables

About this topic

There are many applications of linear inequalities. In this course you will be considering inequalities in two variables; these can be illustrated graphically on a coordinate grid.

This allows you to find the region of points that satisfy a number of inequalities, and then to find the point within the region that gives the maximum or minimum value of a linear function of the two variables.

This is an example of linear programming. When extended to three or more variables it is carried out on a computer but that is outside the scope of this course, which is limited to two variables and working with graphs.

Before you start, remember ...

- how to plot a straight line.

7.1 Linear inequalities in two variables

REVISED

Key facts

1. **Conventions**
   - The line $ax + by = c$ is drawn broken if $ax + by < c$.
   - The line $ax + by = c$ is drawn continuous if $ax + by \leqslant c$.
   - It is usual to shade the side of the line that is not required. This is the convention that has been adopted in this course.
2. **Setting up the inequalities**
   - When a problem is given in context, start by defining the variables.
   - Then express the information given in the form of inequalities.
   - Then plot the inequalities.
3. **The feasible region**

   The set of points which satisfy all the inequalities is called the **feasible region**. This is the unshaded region on a graph when the above shading convention is used.
4. **The objective function**

   The **objective function** is a linear expression in the two variables (often x and y) for which the maximum or minimum value is required.

If it is not obvious which side of the line is required, then use a test point. This can often be the origin (0, 0).

This convention ensures that the region you are working with remains clear and does not get covered by shading.

Worked example

Creating the inequalities

1 A delivery firm has a number of large and a number of small vans. A large van can carry 400 parcels and a small van can carry 200 parcels.

One day the firm must deliver 2000 parcels to an industrial estate.
(i) Write this need as an inequality.
The firm need to use at least as many large vans as small vans.
(ii) Write this condition as an inequality.
(iii) Plot these inequalities on the same graph.

Solution

(i) Let x be the number of large vans and y be the number of small vans used for this delivery.

The problem involves the number of each type of van. So let the number of each be x and y.

$400x + 200y \geqslant 2000$

The left-hand side is the number of parcels that can be delivered using x large vans and y small vans.

Any number of vans that will exceed the need is a possible solution and so this is a greater than or equal to inequality.

The right-hand side is the number of parcels to be delivered.

(ii) 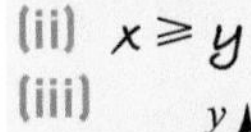
$x \geqslant y$

This again is a greater than or equals to inequality.

(iii)

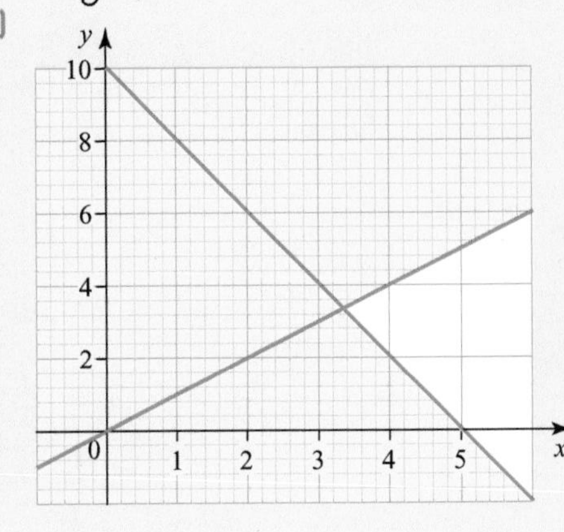

Worked examples

Feasible region

2 Illustrate the feasible region for the following inequalities.

$3 \leqslant x \leqslant 6$

$y \geqslant 0$

$4x + 5y \leqslant 40$

These two inequalities are represented by two lines parallel to the y-axis.

This inequality is represented by the x-axis.

This inequality is represented by an oblique line with negative gradient.

Solution

This is the illustration showing the feasible region.

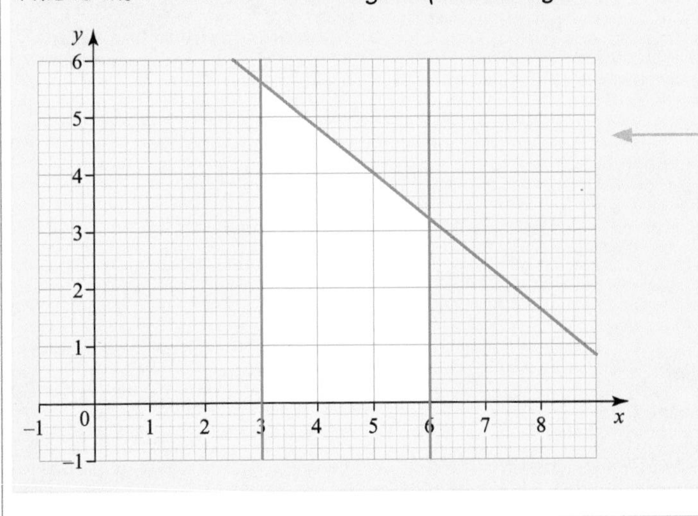

Note that the regions not wanted are shaded. The feasible region is the unshaded part of the grid.

3 You are given the following inequalities.

$y \geqslant 0$

$x \geqslant 1$

$5x + 4y \leqslant 20$

(i) Plot the inequalities on the same grid and shade the regions **not** required.

(ii) List the integer points that lies within the feasible region.

The first inequality is the x-axis.

The second is the line parallel to the y-axis through (1, 0).

The third inequality is represented by an oblique line with negative gradient.

Solution

(i)

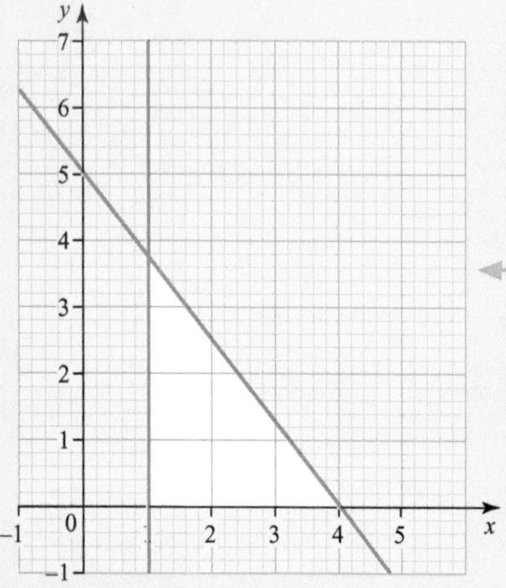

Shading the regions not required gives a triangle as shown.

(ii) Since x and y are required to be integers, the points in the feasible region are:

on the boundaries (1, 0), (1, 1), (1, 2), (1, 3), (2, 0), (3, 0), (4, 0)

inside the region (2, 1), (2, 2), (3, 1)

so there are 10 points in total.

Worked examples

Maximising the objective function

4 Maximise $x + 2y$ given the conditions in Example 3.

Solution

The graph here is the same as that in Example 3 but the line $x + 2y = 5$ has been added.

Sliding this line, parallel to itself, allows you to find other values of $x + 2y$. You will find a ruler helpful.

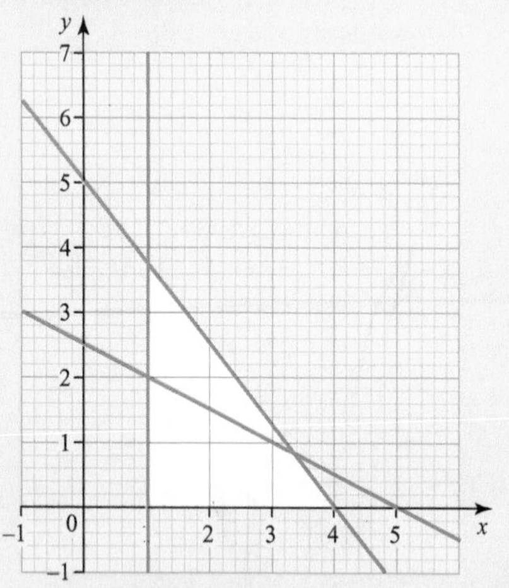

$x + 2y = 5$ at the point (1, 2).

However, at the point (1, 3), $x + 2y = 7$ and this is the maximum value of the function if x and y are integers. If x and y can be any rational number, then the maximum value of $x + 2y$ is at the top point. This is where $5x + 4y = 20$ meets $x = 1$ and is the point (1, 3.75) giving the maximum value of $x = 2y$ of 8.5.

5 (i) Illustrate the feasible region for the following inequalities.

$y \geqslant x, x \geqslant 1, 2x + 3y \leqslant 20$

(ii) Within this region maximise the function

$P = 2x + 5y.$

Solution

(i) *The feasible region is illustrated on the graph.*

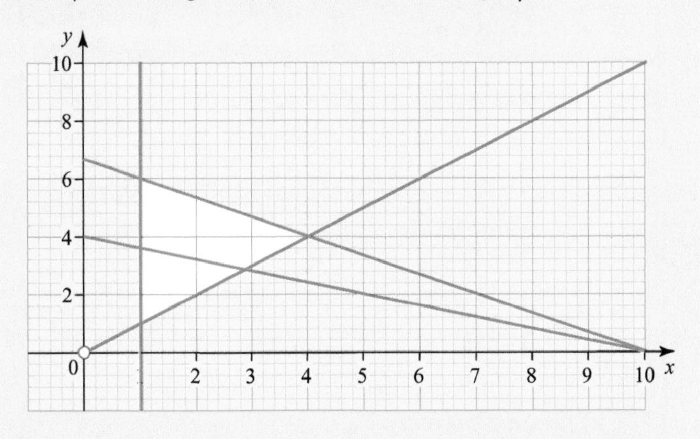

(ii) *The line $2x + 5y = 20$ is shown on the graph. Some points on lines parallel to this line lie within the feasible region.*

In this case the maximum value is at (1, 6), and so

$2x + 5y = 31.$

The problem is to find the maximum value of k where $2x + 5y = k$.

Look for a line through a vertex of the feasible region.

Exam-style question

TESTED

The cost of raw materials plus labour to make a chair is £45 and to make a table is £50.

A furniture maker can only outlay £700 per day.

He must make at least 3 times as many chairs as tables.

In order to meet an expected demand, he must make at least one table.

(i) Express these conditions as three inequalities.

(ii) Illustrate these inequalities on a graph.

The profit made on a chair is £20 and on a table is £35.

(iii) Find the number of chairs and tables that can be made, assuming that they will all be sold, to maximise the profit.

Short answer on page 122

Full worked solution online

CHECKED ANSWERS

Review questions (Chapters 5–7)

1 Find the equation of the line which passes through the points (1, −3) and (2, 6). **[3]**

2 The vertices of a quadrilateral are A (−2, 0), B (2, 2), C (7, −3) and D (0, −4).

(i) Calculate the gradients of the diagonals AC and BD and state a geometrical fact about these lines. **[3]**

(ii) Show that the mid-point of BD lies on AC. **[3]**

3 (i) Show that the two lines whose equations are given below are parallel.

$y = 4 - 2x$

$4x + 2y = 5.$ **[2]**

(ii) Find the equation of the line which is perpendicular to these two lines and which passes through the point (1, 6). **[3]**

4 (i) A circle has equation $x^2 + y^2 - 2x - 4y - 20 = 0$.

Find the coordinates of its centre, C, and its radius. **[3]**

(ii) Find the coordinates of the points where the line $y = x + 2$ cuts the circle. **[5]**

5 The points A and B have coordinates (1, 2) and (5, 8) respectively.

(i) Find the equation of the line AB. **[3]**

(ii) C is the midpoint of AB. Find the coordinates of C. **[1]**

(iii) D is a point on the line through C which is perpendicular to AB and which lies on the y-axis. Find coordinates of D. **[4]**

(iv) Find the equation of the circle which has D as its centre and for which AB is a tangent. **[3]**

6 (i) Find the centre and radius of the circle with equation $x^2 + y^2 - 4x - 6y - 12 = 0$. **[3]**

(ii) Show that the point T (5, 7) lies on the circle. **[1]**

(iii) Find the equation of the tangent to the circle at T. **[4]**

(iv) Find the equation of the other tangent to the circle which is parallel to the line in (iii). **[4]**

7 On the grid is drawn a line.

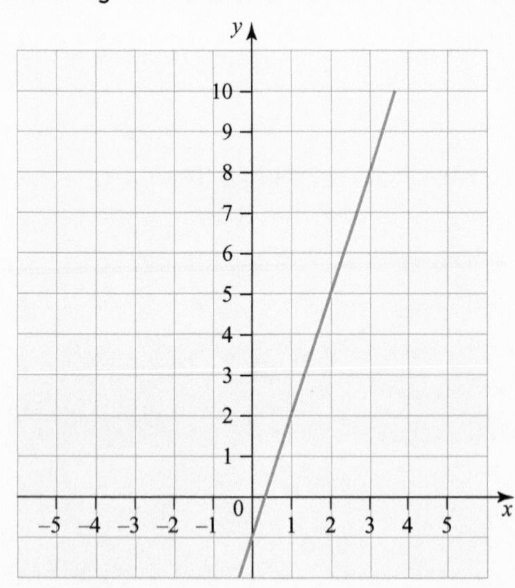

(i) On the same grid draw the line with equation $2x + 3y = 8$. **[2]**

(ii) Write down the coordinates of the point where the two lines meet. **[1]**

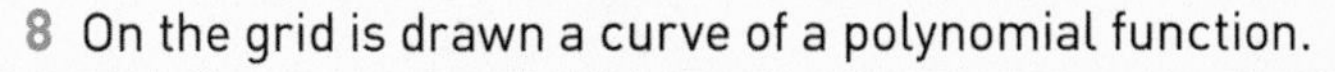

8 On the grid is drawn a curve of a polynomial function.

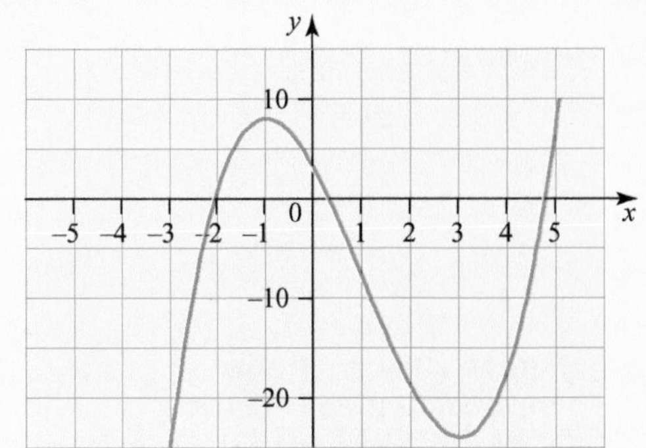

(i) Explain why it is likely that the function is a cubic function. **[1]**

(ii) Estimate the coordinates of the two turning points. **[2]**

(iii) Estimate the coordinates of the points where the gradient of the tangent is 15. **[2]**

9 (i) Plot on the same graph the curves $y = \sin\theta$ and $y = \cos(\theta - 30)$ for $0 \leqslant \theta \leqslant 180°$. **[4]**

(ii) Hence give approximate values of θ that satisfy the equation $\sin\theta = \cos(\theta - 30)$. **[2]**

10 Mikka has collected some data and believes that they fit the exponential function $y = ka^x$.

Mikka proposes that $a = 2$.

x	0	0.5	1.0	1.5	2
y	3	0.7	7.5	12	19

(i) Explain why $k = 3$. **[1]**

(ii) Plot the curve $y = 3 \times 2^x$ on the same grid as the data in the table to show that Mikka is not correct. **[4]**

(iii) If the data fit the curve $y = 3 \times a^x$, then what can you say about the value of a? **[1]**

11 (i) By drawing suitable graphs on the same axes, indicate the region for which the following inequalities hold. You should shade the region which is **not** required.

$3x + 4y \leqslant 24$

$3x + y \leqslant 15$

$x > 0$

$y > 0$ **[4]**

(ii) Find the maximum value of $x + 2y$ in the region for integer values of x and y. **[2]**

12 A small factory makes two types of components, X and Y. Each component of type X requires materials costing £18 and each component of type Y requires materials costing £11. In each week materials worth £200 are available.

Each component of type X takes 7 man-hours to finish and each component of type Y takes 6 man-hours to finish. There are 84 man-hours available each week.

Components cannot be left part-finished at the end of the week. In addition, in order to satisfy customer demands, at least two of each type must be made each week.

(i) The factory produces x components of type X and y components of type Y each week. Write down four inequalities for x and y. **[4]**

(ii) On a graph, draw suitable lines and shade the region that the inequalities do not allow. (Take 1 cm = 1 component on each axis.) **[5]**

(iii) If all components made are sold and the profit on each component of type X is £70 and on each component of type Y is £50, find from your graph the optimal number of each that should be made and the total profit per week. **[3]**

Short answers on pages 122–24

Full worked solutions online

CHECKED ANSWERS

Section 3 Trigonometry

Target your revision (Chapters 8–9)

Try answering each question below. If you get stuck, follow the page reference underneath to revise that topic.

1 **Angles greater than 90°**
Write down all the values of θ in the range $0° \leqslant \theta \leqslant 360°$ that satisfy $\sin\theta = 0.4$, giving your answers correct to 1 decimal place.
(see page 46)

2 **Graphs of the trigonometrical ratios**
Sketch the graph of $y = 1 + \sin\theta$ for values of θ in the range $0° \leqslant \theta \leqslant 360°$.
(see page 46)

3 **Area of a triangle**
In the triangle ABC, AB = 4 cm, BC = 5 cm and angle B = 70°. Find the area of the triangle.
(see page 49)

4 **Cosine rule**
In the triangle ABC, AB = 4 cm, BC = 5 cm and angle ABC = 40°.
Find the length of AC.
(see page 49)

5 **Sine rule**
In the triangle PQR, angle P = 55°, angle R = 65° and PR = 7 cm.
Find the lengths of PQ and RQ.
(see page 49)

6 **The ambiguous case for the sine rule**
The triangle ABC has the angle at B = 40°.
AB = 6 cm and AC = 5 cm.
Show that there are two possible angles for C and find their values.
(see page 49)

7 **Identities**
Prove that $\dfrac{\cos\theta + \sin\theta}{\cos\theta} = 1 + \tan\theta$.
(see page 51)

8 **Trigonometrical equations**
Solve the equation $\sin\theta = 2\cos\theta$.
(see page 51)

9 **Applications of trigonometry – line of greatest slope**
A sloping rectangular field, ABCD, can be modelled as the top face of a wedge, as shown in the diagram.

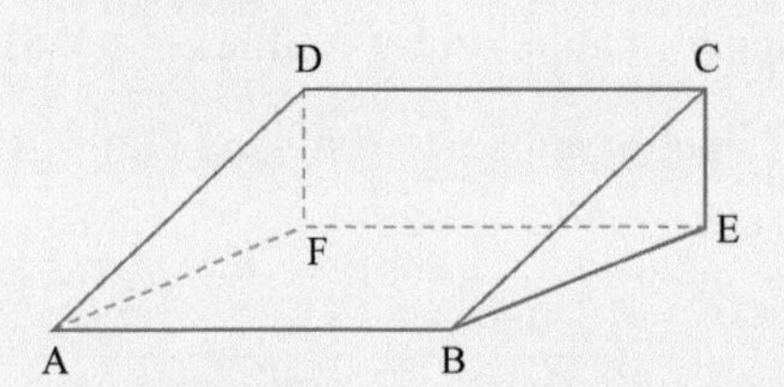

The horizontal base of the wedge, ABEF, is such that E is vertically below C and F is vertically below D.
The vertical face, ECDF, is also a rectangle.

AB = CD = EF = 100,

BE = AF = 50,

CE = DF = 20.

Units are metres.
Find the angle of greatest slope of the field.
(see page 53)

10 **Angle between a line and a plane**
In the example in question 9, find the angle of slope of a path which runs from A to C.
(see page 53)

Short answers on page 124
Full worked solutions online

CHECKED ANSWERS

Chapter 8 Trigonometric functions

About this topic

You have already met the trigonometric ratios sin, cos and tan, for angles between 0° and 90°, and used them to solve right angled triangles. In this chapter the definition is extended to include angles of any size, and you learn to solve any triangle. Working with angles of any size allows you to think of $\sin\theta$, $\cos\theta$ and $\tan\theta$ as mathematical functions in their own right. This important step is illustrated by the final topic on identities.

Before you start, remember ...

- the three trigonometric ratios: sin, cos and tan.

8.1 Angles greater than 90°

REVISED

Key facts

1 **Trigonometric ratios for angles of any size**
P is a general point (x, y). The distance OP is r and the angle OP makes with the x-axis is θ in the anticlockwise direction. The trigonometric ratios are defined as follows:

$$\sin\theta = \frac{y}{r},\ \cos\theta = \frac{x}{r},\ \tan\theta = \frac{y}{x}$$

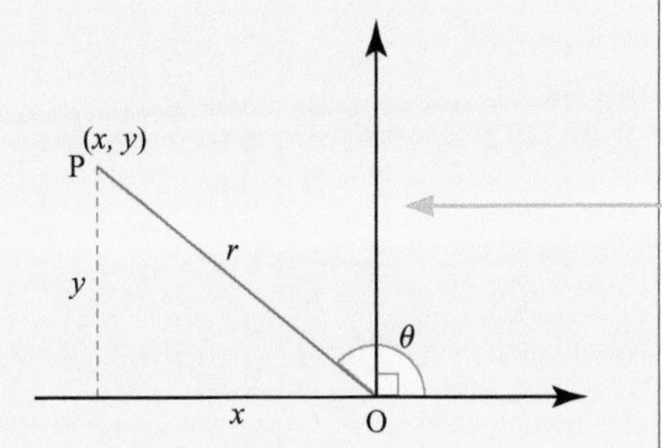

2 **Quadrants**
The axes define four quadrants as shown below.

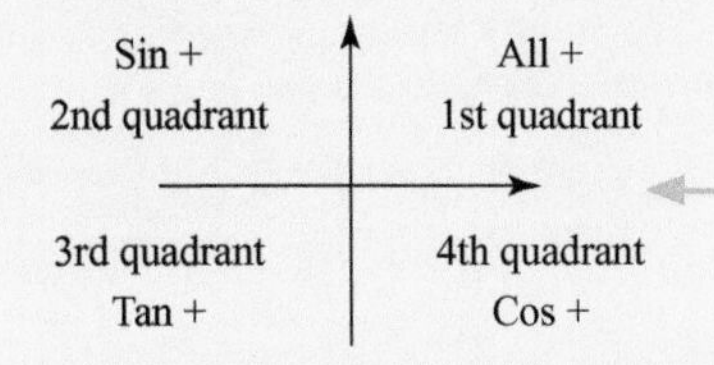

In the first quadrant all three ratios are positive.
In the second quadrant (angles between 90°and 180°), the sine of an angle is positive and the other two ratios are negative.
And so on.

3 **Finding angles**
There are two conventions for an angle with a given trig ratio. If, for example, $\sin\theta = 0.7$, then θ can be written as either $\sin^{-1} 0.7$ or as $\arcsin 0.7$.

For any possible value of a ratio there are two angles in the range $0° \leqslant \theta \leqslant 360°$.

This is consistent with the definitions for a right-angled triangle in that the definitions hold for the first quadrant. These definitions give the ratio for any angle.

Note: You will sometimes meet the angle labelled x rather than θ although x is usually reserved for a length.

One way to remember in which quadrant ratios are positive and negative is the mnemonic CAST.

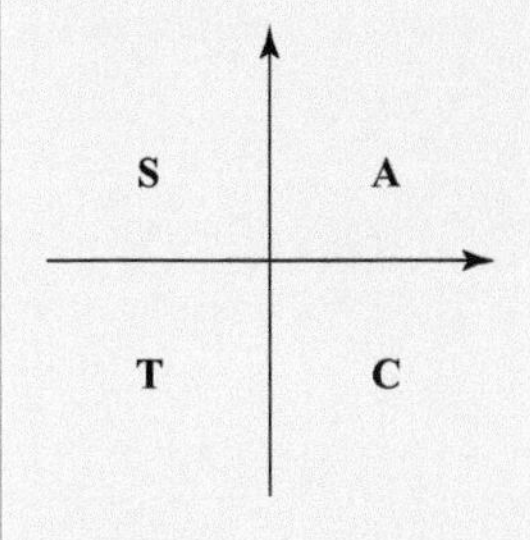

Example: $\cos 100° = -0.174$, $\sin 100° = 0.985$, $\tan 100° = -5.671$.

4 Principal angles

When you use your calculator to find an angle, only one of the two angles in the range [0°, 360°] is shown. The angle shown on your calculator is called the principal angle and the other one has to be worked out.
Principal angles lie in the following ranges:

$\cos\theta$: $0° < \theta < 180°$,

$\sin\theta$ and $\tan\theta$: $-90° < \theta < 90°$.

5 Graphs of trigonometric functions

The graphs of the trigonometric functions have distinctive shapes.

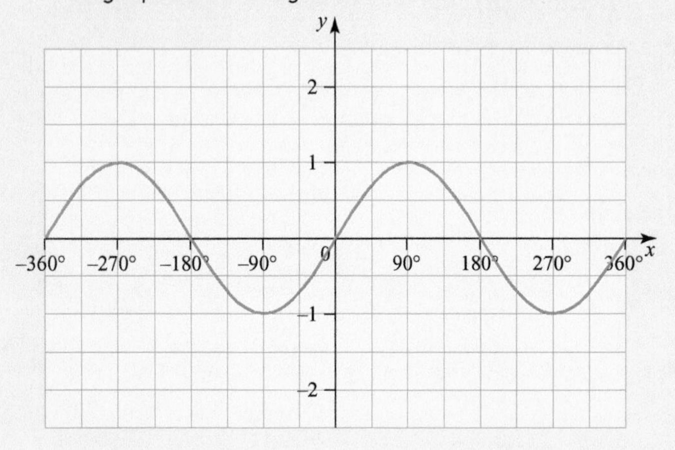

You need to know these curves and to be able to use them in problem solving.

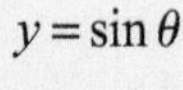

$y = \sin\theta$

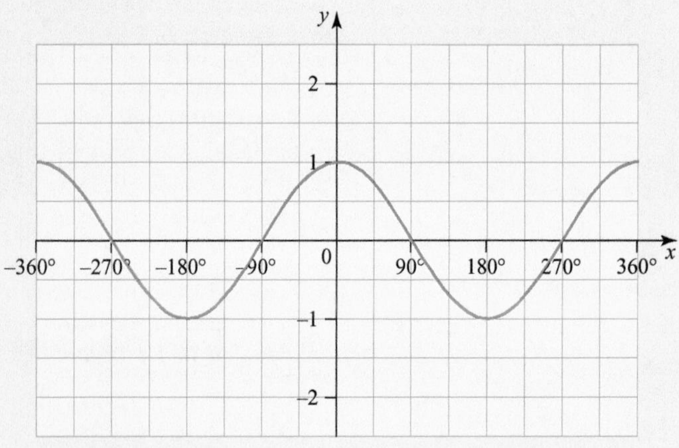

$y = \cos\theta$

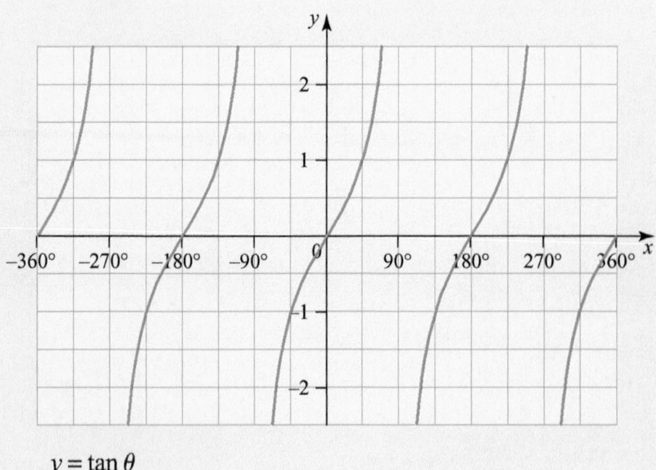

$y = \tan\theta$

Worked examples

Angles greater than 90°

1 You are given that $\cos\theta = 0.4$.
Find the two values of θ in the range $0° \leqslant \theta \leqslant 360°$.

Solution

The principal angle is $\theta = 66.4°$.

The other angle is in the fourth quadrant

and so is $\theta = 360° - 66.4° = 293.6°$.

2 Find the two values of θ in the range $0° \leqslant \theta \leqslant 360°$ that satisfy $\tan\theta = -0.3$.

Solution

Using a calculator, $\theta = -16.7°$ and this is the principal angle.

The principal angle is in the range $-180° < \theta < 180°$.

This is outside the required range.

The two values are $360° - 16.7° = 343.3°$

and $180° - 16.7° = 163.3°$.

A sketch of $y = \tan\theta$ and $y = -0.3$ will show you that the tan ratio is negative in the second and fourth quadrants.

Worked example

Sketching the graphs of the trigonometric ratios

3 Sketch the curve $y = \cos 2\theta$.

Solution

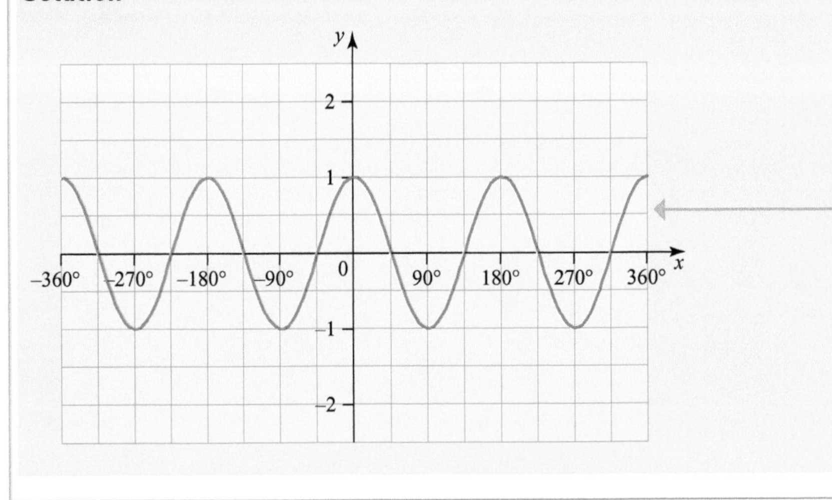

The basic shape is a cosine graph which has maximum and minimum values of 1 and −1. Because of the '2', the graph repeats itself after 180°.

Exam-style question

TESTED

(i) On the same grid, sketch the graphs of the curves $y = \sin(2\theta + 60)$ and $y = 2\cos\theta$ in the range $0° \leqslant \theta \leqslant 360°$.

(ii) From the graph estimate the value of θ that satisfies the equation $\sin(2\theta + 60) = 2\cos\theta$ in the range $90° \leqslant \theta \leqslant 180°$.

Short answers on page 124

Full worked solutions online

CHECKED ANSWERS

8.2 Sine and cosine rules

REVISED

Key facts

1 **Convention**

The three sides of a triangle are labelled a, b and c with side a opposite vertex A, etc. The angle at vertex A is A and so on.

This convention will be used throughout this book and is used in the examinations.

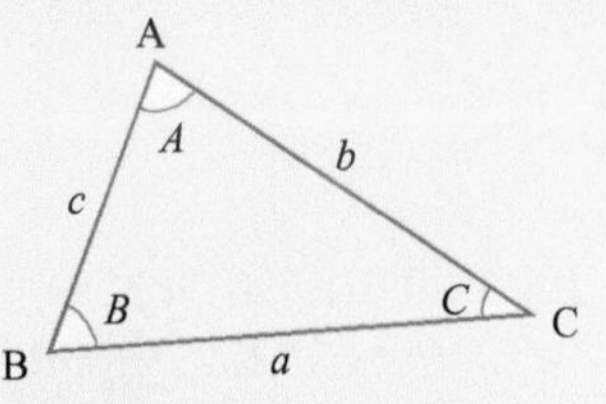

2 **The sine rule**

In the triangle ABC,

$\dfrac{a}{\sin A} = \dfrac{b}{\sin B} = \dfrac{c}{\sin C}$ which can be written $\dfrac{\sin A}{a} = \dfrac{\sin B}{b} = \dfrac{\sin C}{c}$.

3 **The cosine rule**

In the triangle ABC,

$a^2 = b^2 + c^2 - 2bc\cos A$.

This can be written in two other ways:

$b^2 = c^2 + a^2 - 2ca\cos B$ and $c^2 = a^2 + b^2 - 2ab\cos C$.

The angle can also be made the subject of the formula:

$\cos A = \dfrac{b^2 + c^2 - a^2}{2bc}$.

4 **Area**

Note: The area of the triangle is assumed knowledge in this course.

$$\text{Area of triangle} = \frac{1}{2} \times \text{base} \times \text{height}$$

$$= \frac{1}{2}ah$$

$$h = b\sin C$$

$$\Rightarrow \quad \text{Area} = \frac{1}{2}ab\sin C$$

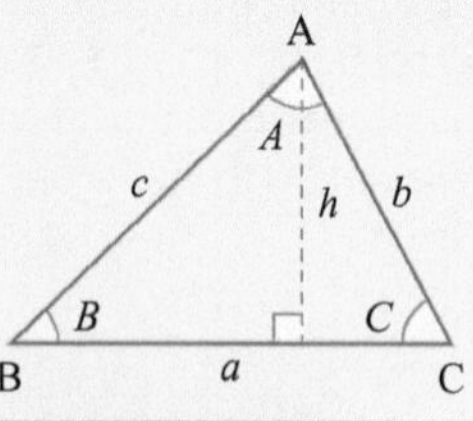

Worked example

The sine rule

1 In a triangle $a = 5$, $A = 56°$ and $B = 47°$. Find b.

Choose the appropriate part of the sine rule and substitute.

Solution

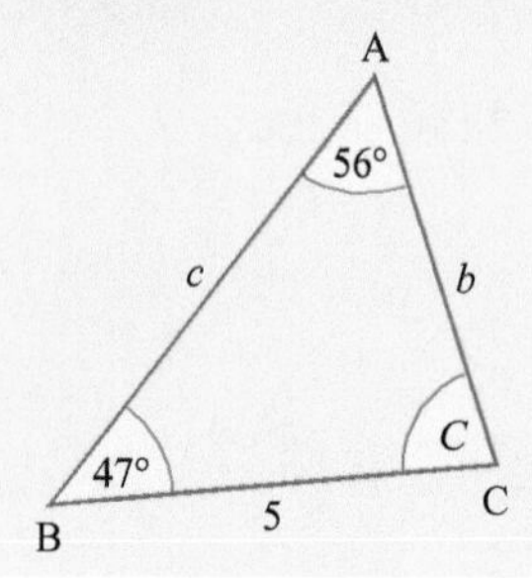

Using the sine rule

$$\frac{a}{\sin A} = \frac{b}{\sin B}$$

$$\frac{5}{\sin 56} = \frac{b}{\sin 47}$$

$$\Rightarrow \quad b = \frac{5\sin 47}{\sin 56} = 5.77$$

Make b the subject and solve.

Worked example

The ambiguous case

2 In triangle PQR, $r = 10$ cm, $q = 7$ cm and $Q = 40°$.

(i) Show that there are two possible values for the triangle R and find the two values.

(ii) Illustrate the two triangles on a diagram.

Solution

(i) $\frac{\sin R}{r} = \frac{\sin Q}{q} \Rightarrow \sin R = \frac{10\sin 40}{7} = 0.9182$

There are two values of R in the range $0° \leqslant R \leqslant 180°$ which satisfy the equation $\sin R = 0.9182$

$\Rightarrow R = 66.7°$

But also $R = 180 - 66.7 = 113.3°$

(ii)

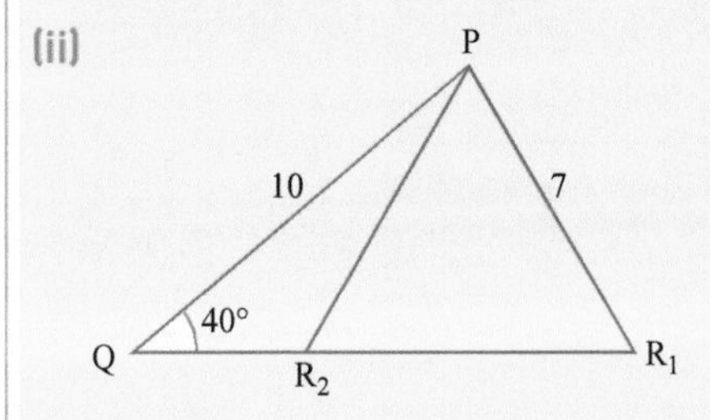

Note that the triangle PR_1R_2 is isosceles with angle PR_2R_1 = PR_1R_2. The angle QR_2P and angle PR_2R_1 are on a straight line and so add to 180°

Worked examples

The cosine rule

3 In a triangle, $A = 37°$, $b = 4$ and $c = 6$. Find a.

Solution

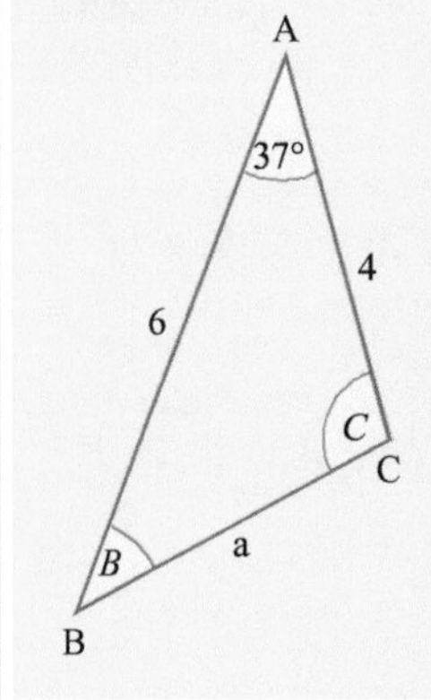

Using the cosine rule

$$a^2 = b^2 + c^2 - 2bc\cos A$$

$$a^2 = 4^2 + 6^2 - 2 \times 4 \times 6\cos 37$$

$$= 13.67$$

$$\Rightarrow a = 3.70$$

Common mistake: Take care to calculate the third term before subtracting from the sum of the first two terms.

4 The three sides of a triangle are 6, 7 and 8. Find the size of the largest angle.

Solution

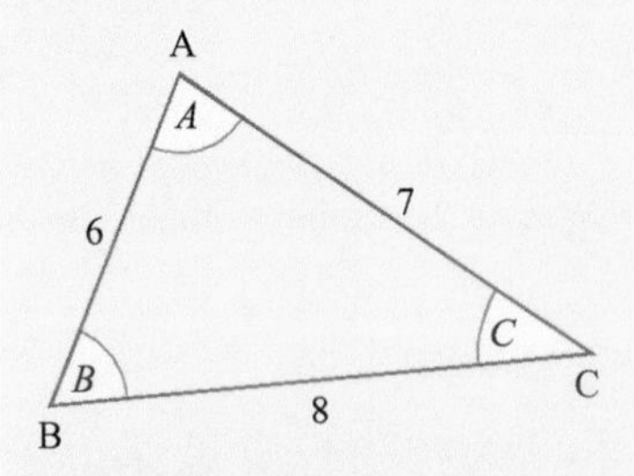

The largest angle is opposite the largest side which is 8

using the cosine rule

$$\cos A = \frac{b^2 + c^2 - a^2}{2bc}$$

$$\cos A = \frac{6^2 + 7^2 - 8^2}{2 \times 6 \times 7} = 0.25$$

$$\Rightarrow A = 75.5^\circ$$

Exam-style question

TESTED

Adam and Beth set out walking from a point P at the same time. After one hour, Adam is at A, 3.6 km due North of P, and Beth is at B, 2.5 km from P on a bearing of 035°.

Calculate how far they are apart when they have walked for one hour. Give your answer correct to 2 significant figures.

Short answer on page 124

Full worked solution online

CHECKED ANSWER

8.3 Identities and related equations

REVISED

Key facts

1 **Identities and equations**
- An **identity** is true for all permissible values of the variable. The symbol ≡ is sometimes used.
- By contrast an **equation** is only true for certain values. The symbol = is used.

2 **Trigonometric equations**
Simple trigonometric identities are often used to simplify and solve more complicated equations.
Drawing graphs is helpful in determining the solution to trigonometric equations.

For example: $\tan^2\theta + 1 \equiv \frac{1}{\cos^2\theta}$ is true for all values of θ except 90°, 270°, ... when $\tan\theta$ is not defined. It is an identity.
$x^2 - 3x - 40 = 0$ is an equation. It is true for only two values of x, 8 and –5.

For example: the equation $2\sin^2\theta + \cos\theta = 2$ can be solved by using the identity $\sin^2\theta + \cos^2\theta = 1$ to make it into a quadratic equation in $\cos\theta$.

A sketch of the functions gives you an indication of all possible answers.

Worked examples

Solution of equations

1 Solve the equation $\sin\theta = 2\cos\theta$ for $0° \leqslant \theta \leqslant 360°$.

Solution

$\sin\theta = 2\cos\theta$

$\Rightarrow \dfrac{\sin\theta}{\cos\theta} = 2$

$\Rightarrow \tan\theta = 2$

$\Rightarrow \theta = 63.4°$ or $243.4°$

Divide by $\cos\theta$.

Using the identity $\tan\theta = \dfrac{\sin\theta}{\cos\theta}$.

2 Solve the equation $2\cos^2\theta + \sin\theta = 1$ for $0° \leqslant \theta \leqslant 360°$.

Solution

$2\cos^2\theta + \sin\theta = 1$

$\Rightarrow 2(1 - \sin^2\theta) + \sin\theta = 1$

$\Rightarrow 2 - 2\sin^2\theta + \sin\theta = 1$

$\Rightarrow 2\sin^2\theta - \sin\theta - 1 = 0$

$\Rightarrow (2\sin\theta + 1)(\sin\theta - 1) = 0$

$\Rightarrow \sin\theta = 1$ or $-\dfrac{1}{2}$

$\Rightarrow \theta = 90°, 210°$ or $330°$

Use the identity $\cos^2\theta = 1 - \sin^2\theta$ and arrange into a quadratic equation in $\sin\theta$.

Solve by factorising.

Your calculator will probably show you that $\theta = -30°$, which is not within the range required. If the value of $\sin\theta$ is negative then θ is in the third and fourth quadrants.

Exam-style question

TESTED

It is required to solve the equation $\sin\theta\cos\theta = \dfrac{1}{4}$.

(i) Show that $\dfrac{\sin\theta}{\cos\theta} + \dfrac{\cos\theta}{\sin\theta} = \dfrac{1}{\sin\theta\cos\theta}$.

(ii) Hence show that the equation $\sin\theta\cos\theta = \dfrac{1}{4}$ is equivalent to $\tan\theta + \dfrac{1}{\tan\theta} = 4$.

(iii) By expressing this as a quadratic equation in t where $t = \tan\theta$ find two values of θ in the range $0° \leqslant \theta \leqslant 180°$ that satisfy the equation.

Short answer on page 124

Full worked solution online

CHECKED ANSWER

Chapter 9 Applications of trigonometry

About this topic

The techniques learned in the previous chapter can be used in practical problems in 2 and 3 dimensions.

Before you start, remember ...

- the three trigonometric ratios, $\sin\theta$, $\cos\theta$, and $\tan\theta$, for any angle
- the sine and cosine rules.

9.1 Applications of trigonometry

REVISED

Key facts

1 **Choices of rules for solving triangles**

You may have to decide which formula to use.

Given	To be found	Rule
SAS	Third side	Cosine rule
SSS	Any angle	Cosine rule
AAS	Any side	Sine rule
ASS	Third side or either angle	Sine rule, but beware the ambiguous case

If you know two angles of a triangle you also know the third because their total is 180°.

2 **Working in three dimensions**

True shape diagrams

Triangles are two-dimensional shapes.

Drawings of three-dimensional objects are always distorted; for example, right angles do not always appear to be 90°. Always start by drawing true shape diagrams of any triangles with which you are going to work.

3 **The angle between a line and a plane**

This diagram shows a line AB crossing a plane at A. The point C is also on the plane and angle ACB = 90°. The angle between the line and the plane is BAC.

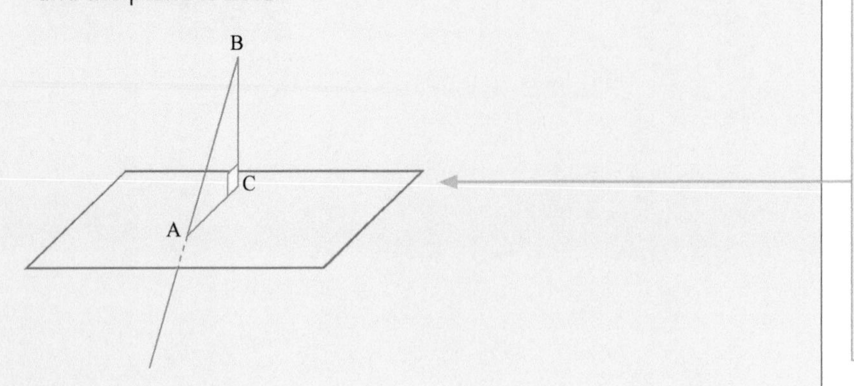

This is the true shape diagram of triangle ABC.

4 Angle of greatest slope

The diagram shows a sloping plane meeting the horizontal in a straight line. The line AB is perpendicular to that line and lies in the sloping plane. It is a **line of greatest slope** and the angle θ is the **angle of greatest slope**.

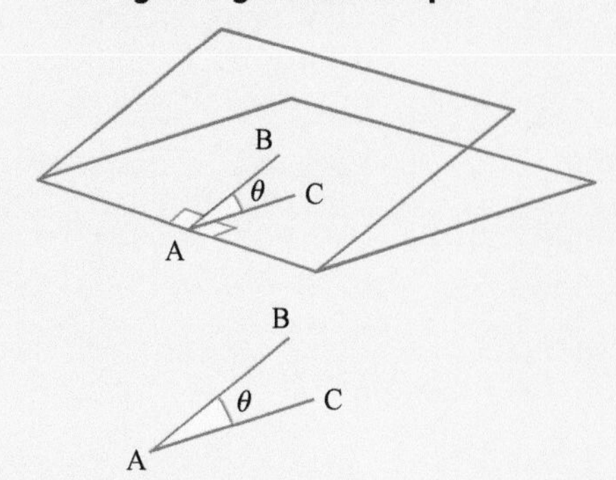

Worked example

Angle between a line and a plane

1 The diagram shows a cuboid with horizontal base ABCD.

AB = 5 cm and BC = 4 cm.

The top EFGH is such that E is vertically above A with AE = 3 cm.

Find the angle that the diagonal AG makes with

(i) the base, ABCD,

(ii) the face ABFE.

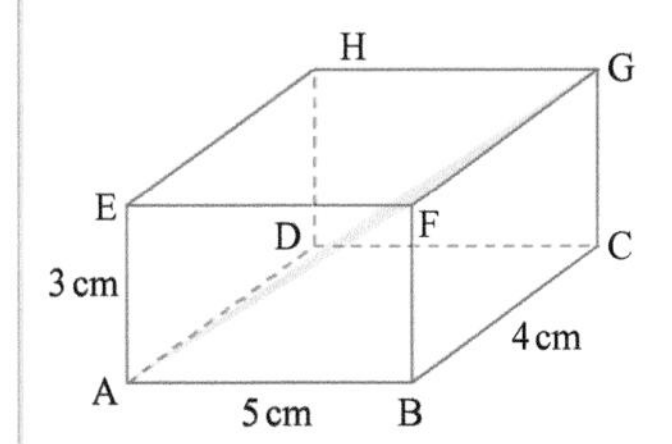

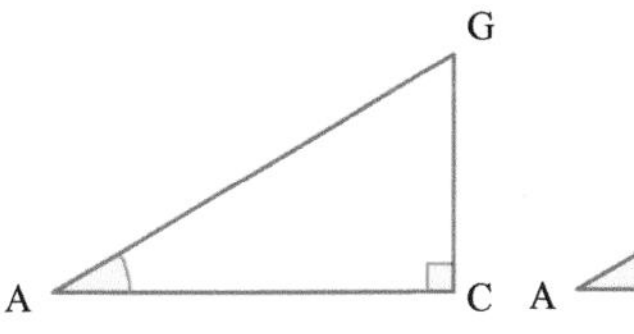

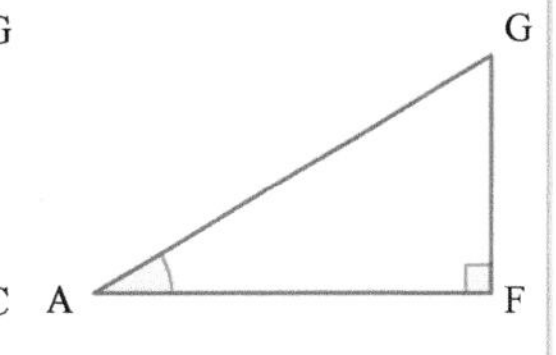

Solution

(i)

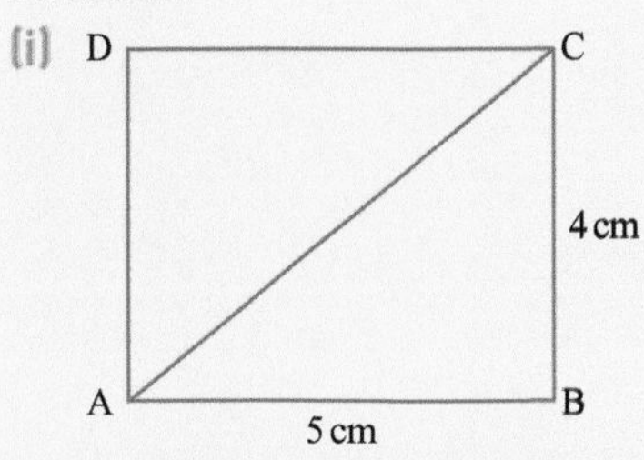

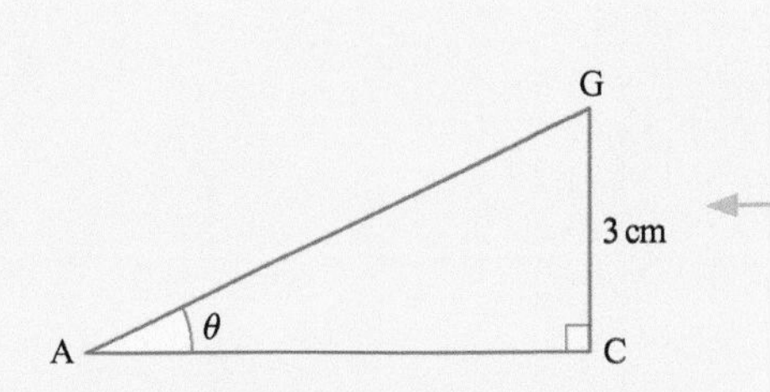

These are true shape diagrams of triangles ABC and ACG. In part (i) the required angle is GAC.

AC is the diagonal of the rectangular base, ABCD.

The required triangle is ACG. The right angle is at C and the required angle, θ, is GAC.

In triangle ABC, $AC^2 = 4^2 + 5^2 = 41$ so $AC = \sqrt{41}$.

In triangle AGC, $\tan\theta = \dfrac{CG}{AC}$.

So $\tan\theta = \dfrac{3}{\sqrt{41}} \Rightarrow \theta = 25.1°$.

(ii)

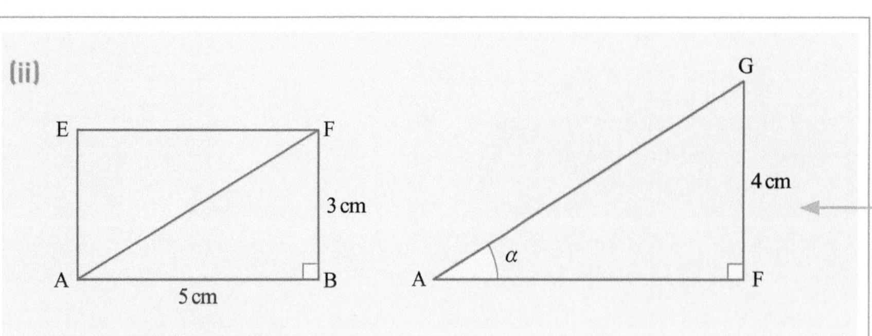

AF is the diagonal of the vertical rectangular face AEFB.

The required angle, α, is GAF.
In triangle ABF, $AF^2 = 5^2 + 3^2 = 34$.

So $\tan\alpha = \frac{4}{\sqrt{34}} \Rightarrow \alpha = 59.1°$.

Worked example

Angle of greatest slope

2 The diagram shows a hillside of uniform slope.

AB = 200 m, BC = 100 m and EC = 30 m.

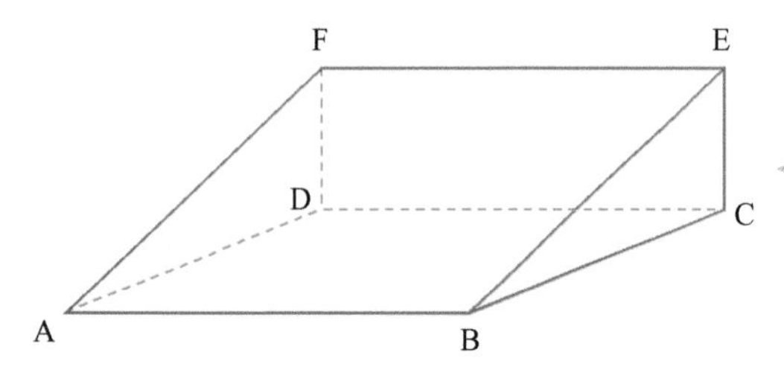

A hillside with uniform slope may be modelled as a wedge as shown in the diagram. ABCD is a horizontal rectangular plane, the hillside is the rectangle ABEF such that FDCE is a vertical rectangular plane.

Find

(i) the line of greatest slope

(ii) the angle between the line AE on the slope and the horizontal.

Solution

(i) *An angle of greatest slope is EBC.*

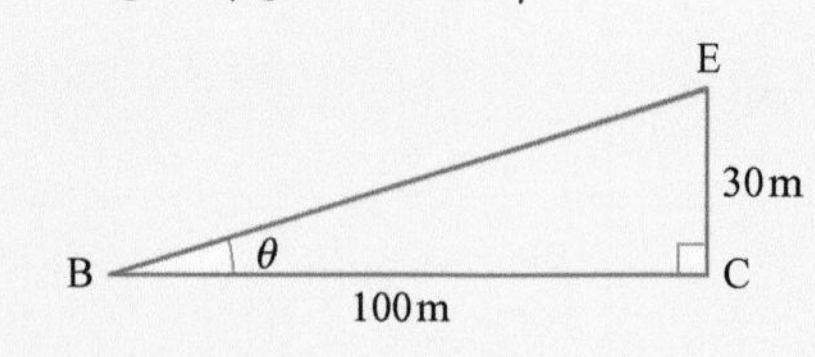

BC and BE are perpendicular to the common line, AB and so EBC is the angle of greatest slope.

Another angle of greatest slope is FAD.

Angle EBC = $\tan^{-1} = \left(\frac{30}{100}\right) = 16.7°$.

(ii) *The triangle to be used is CAE.*

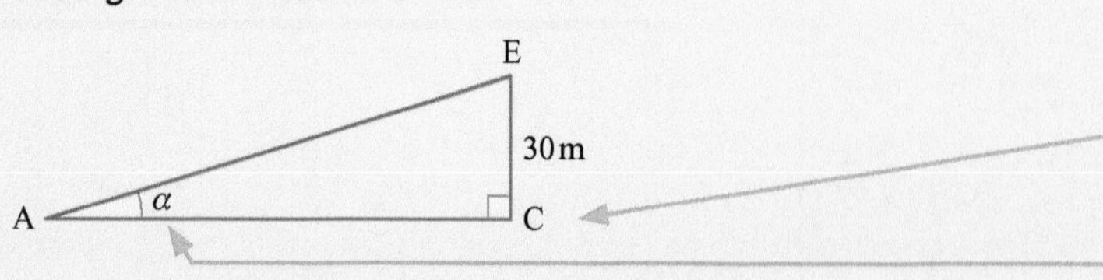

In the triangle EAC the line AC must be found.

The angle that AE makes with the horizontal is angle CAE.

However, first the length of AC must be found.
This is done from the triangle ABC in the rectangle ABCD.

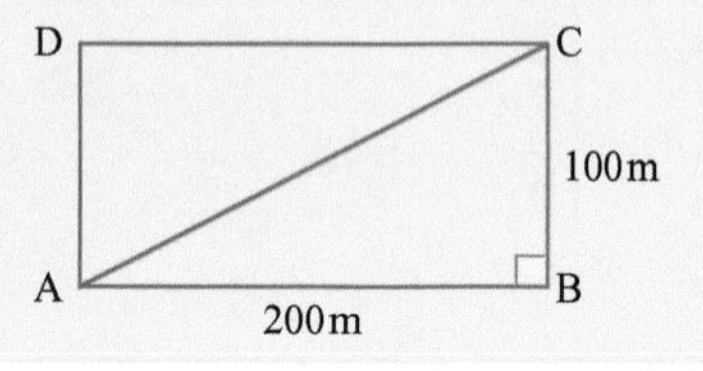

$$AC = \sqrt{AB^2 + BC^2}$$

$$= \sqrt{200^2 + 100^2} = 223.6...$$

Keep the unrounded number on your calculator until the very end.

In triangle CAE

$$\text{Angle } CAE = \tan^{-1}\left(\frac{EC}{AC}\right)$$

$$\Rightarrow \text{Angle } CAE = \tan^{-1}\left(\frac{30}{223.6...}\right) = 7.6° \text{ (to 1 dp)}$$

Exam-style question

TESTED

A pyramid ABCDV has a square horizontal base of side 6 cm. The vertex V is vertically above the centre of the base, O. The pyramid has height 7 cm.

Find the angle that the sloping edge VA makes with the horizontal.

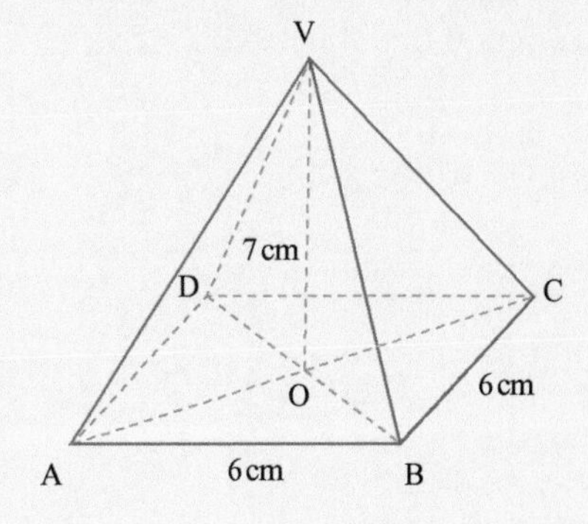

Short answer on page 124

Full worked solution online

CHECKED ANSWER

Review questions (Chapters 8–9)

1 Find the four values of x in the range $0° \leqslant x \leqslant 360°$ that satisfy the equation $\sin 2x = 0.5$. **[4]**

2 Find the value of x in the range $0° < x < 360°$ that satisfies **both** $\tan x = 0.75$ **and** $\cos x = -0.8$. **[3]**

3 Find all the values of x in the range $0° < x < 360°$ that satisfy $\sin x = -2\cos x$. **[3]**

4 A triangle ABC has sides AB = 6 cm and CB = 8 cm as shown in the diagram.

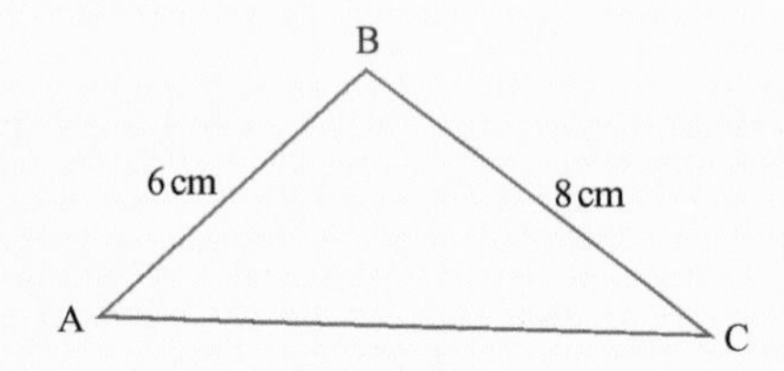

(i) When angle ABC = 90° find the area of the triangle. **[2]**

(ii) When ABC ≠ 90° you are given that the area is 20 cm². Find the two possible angles ABC. **[4]**

5 A triangular wedge of land ABC is such that AB = 10 m, BC = 12 m and the angle ABC = 50°. Find the length of the side AC. **[3]**

6 Solve the equation $\sin^2\theta = 12 - 13\cos\theta$ for values of θ in the range $0° \leqslant \theta \leqslant 360°$. **[4]**

7 Use the given triangle to prove that, for $0° < \theta < 90°$, $1 + \tan^2\theta = \dfrac{1}{\cos^2\theta}$. **[3]**

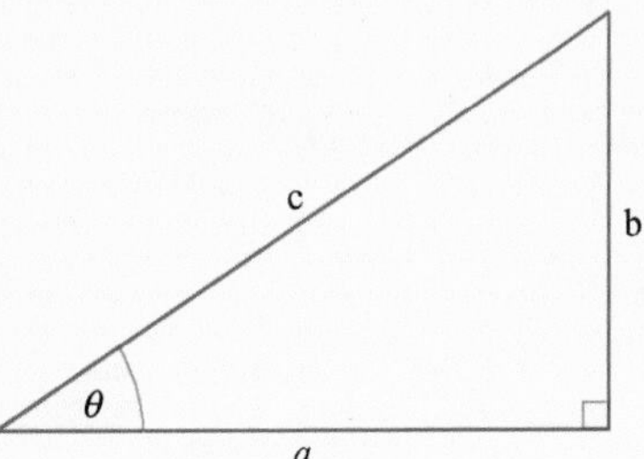

8 A pyramid stands on a horizontal triangular base, ABC, as shown in the diagram.
The angles CAB and ABC are 50° and 60° respectively.
The vertex, V, is directly above C with VC = 10 m.
The angle which the edge VA makes with the vertical is 40°.

(i) Calculate AC. **[2]**

(ii) Hence calculate AB. **[4]**

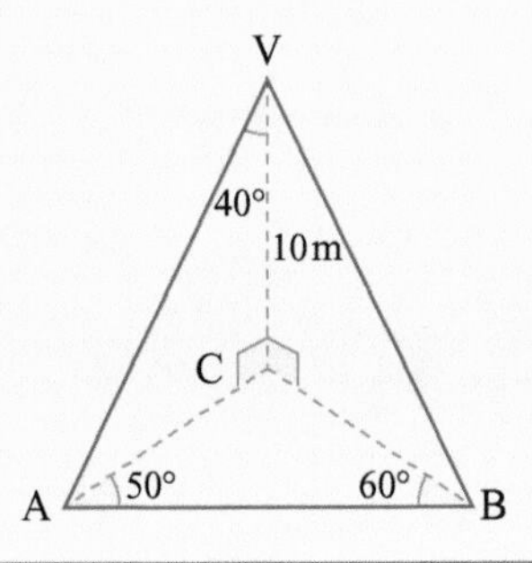

9 The diagram shows a rectangle ABEF on a plane hillside which slopes at an angle of 30° to the horizontal. ABCD is a horizontal rectangle. E and F are 100 m vertically above C and D respectively. AB = DC = FE = 500 m.

AE is a straight path.

From B there is a path which runs at right angles to AE, meeting it at G. H is vertically below G and lies on AC.

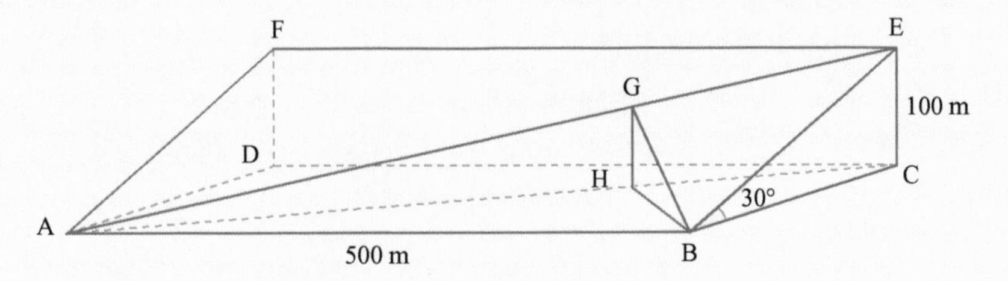

(i) Find the distance BE. **[3]**

(ii) Find the angle that the path AE makes with the horizontal. **[4]**

(iii) Find the area of the triangle ABE.

Hence find the length BG. **[5]**

Short answers on pages 124–25

Full worked solutions online

CHECKED ANSWERS

Section 4 Selections

Target your revision (Chapters 10–11)

Try answering each question below. If you get stuck, follow the page reference underneath to revise that topic.

1 **Two-way tables**
A group of 30 students were asked how they got to college on one particular morning.
Some of the results are given in the table below.

	Walked	Brought by car	Came by bus	Total
Boys	5		10	18
Girls				
Total	9		13	30

Complete the table.
(see page 60)

2 **Venn diagrams**
A group of 40 students were asked about the number of siblings in their family.
11 said that they had one or more sisters but no brothers.
10 said they had one or more brothers but no sisters.
9 said they had both sisters and brothers.
How many had no brothers or sisters?
(see page 60)

3 **Tree diagrams and probability**
Ahmed and Brian play a game of tennis one morning and a game of badminton in the afternoon. The probability of Ahmed winning the tennis match is 0.7 and the probability of Brian winning the badminton is 0.6.
Construct a tree diagram to illustrate the outcomes of the two games for Ahmed and work out the probability that Ahmed wins both games.
(see page 60)

4 **Factorials**
Calculate $\frac{7!}{4!}$.
(see page 63)

5 **Permutations**
John has 5 cards on which are written the numbers 1, 2, 3, 4 and 5.
He lays them down on the table making a 5-digit number.
(i) How many different numbers can he make?
(ii) How many of the numbers are even?
(see page 63)

6 **Combinations**
How many ways are there of choosing 3 students from a group of 30?
(see page 63)

7 **Selections and probability**
Four red and four green balls are placed in a bag.
Amy removes a ball and notes its colour and then removes a second ball, noting its colour.
Find the probability that both balls are red if
(i) the first ball is replaced before the second one is drawn
(ii) the first ball is not replaced before the second one is drawn.
(see page 63)

8 **The binomial expansion – Pascal's triangle**
Expand $(2 - 3x)^5$.
(see page 66)

9 **The binomial expansion – calculating coefficients**
Find the first 4 terms in the expansion of $\left(1+\frac{1}{2}x\right)^{10}$.
(see page 66)

10 **The binomial expansion – applications**
Find the constant term in the expansion of $\left(2x-\frac{3}{x}\right)^{6}$.
(see page 66)

11 **The binomial distribution**
A large consignment of components is delivered in boxes of 10.
Of the components, 10% are known to be faulty, with the faults occurring at random and independently of each other.
A box is chosen at random.
(i) What is the probability that there are no faulty components in the box?
(ii) What is the probability that there are 2 or more faulty components in the box?
(see page 67)

Short answers on page 125

Full worked solutions online

CHECKED ANSWERS

Chapter 10 Permutations and combinations

About this topic

For equally likely outcomes the probability of a particular outcome is the number of required outcomes divided by the number of possible outcomes. In order to use this definition, you need to be able to work out the numbers on the top and bottom lines of this fraction. This chapter is about how you do this and so involves selections and arrangements.

When you are arranging a number of objects in a certain way, there are two cases to consider: (i) when order matters and (ii) when order does not matter. These are called permutations and combinations, respectively.

Before you start, remember ...

- the definition of probability.

10.1 Definitions and probability

REVISED

Key facts

1 **Notation and definitions**
In statistics an **experiment** is a means of collecting data; the number of possible **outcomes** may be finite or infinite. The set of all possible outcomes is called the **sample space**. An illustration (or list) of all outcomes is called a **sample space diagram** or a **two-way table**.
An **event** is a specific outcome or set of outcomes.

2 **Diagrams**
There are many ways of showing the outcomes from an experiment including
- a tree diagram
- a Venn diagram
- a sample space diagram
- a two-way table.

3 **Probability notation, language and laws**
$\mathrm{P}(A)$ is the probability that an event A happens.
$\mathrm{P}(A')$ is the probability that event A does not happen so

$$\mathrm{P}(A') = 1 - \mathrm{P}(A).$$

A' is the **complement** of A.
When two events are **independent** then

$$\mathrm{P}(A \text{ and } B) = \mathrm{P}(A) \times \mathrm{P}(B).$$

If only one of two events can occur, they are **mutually exclusive** so

$$\mathrm{P}(A \text{ and } B) = 0.$$

In a Venn diagram, the circles representing mutually exclusive events A and B do not overlap.

For example, when throwing a six-sided die the sample space is 1, 2, 3, 4, 5, 6.

For example, a die showing a 2, or a die showing an even number.

Worked example

Sample space (or two-way table)

1 Two dice are thrown and their scores are added.
 (i) Draw up a two-way table to illustrate the possible outcomes.
 (ii) Construct a table showing the probabilities of the different possible outcomes.

Solution

(i)

		Die 2					
		1	2	3	4	5	6
Die 1	1	2	3	4	5	6	7
	2	3	4	5	6	7	8
	3	4	5	6	7	8	9
	4	5	6	7	8	9	10
	5	6	7	8	9	10	11
	6	7	8	9	10	11	12

(ii)

Outcome	Probability
2	$\frac{1}{36}$
3	$\frac{2}{36}=\frac{1}{18}$
4	$\frac{3}{36}=\frac{1}{12}$
5	$\frac{4}{36}=\frac{1}{9}$
6	$\frac{5}{36}$
7	$\frac{6}{36}=\frac{1}{6}$
8	$\frac{5}{36}$
9	$\frac{4}{36}=\frac{1}{9}$
10	$\frac{3}{36}=\frac{1}{12}$
11	$\frac{2}{36}=\frac{1}{18}$
12	$\frac{1}{36}$

Notice from this example that the sample space is the outcomes from the experiment (when you add the two numbers together you can only get 2 to 12) and so the sum of probabilities is 1.

Worked example

Tree diagrams

2 When John drives to work he passes two sets of traffic lights. He knows, from experience, that the probability of being stopped at each light is 0.7. The two sets of lights work independently of each other.

(i) Draw a tree diagram to illustrate this information

(ii) Write down the sample space of the number of times he is stopped on a particular day.

(iii) Calculate the probability of each outcome.

Solution

(i)

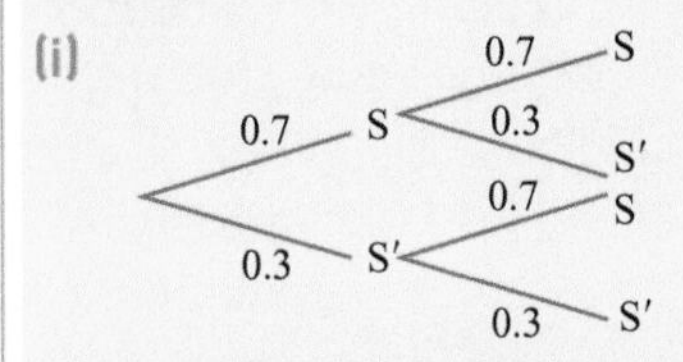

S means stopped, S' means not stopped.

(ii) He can be stopped 0, 1 or 2 times.

(iii)

Outcome	Probability
0	$0.7 \times 0.7 = 0.49$
1	$0.7 \times 0.3 \times 2 = 0.42$
2	$0.3 \times 0.3 = 0.09$

The probability of being stopped at the first set but not the second = $0.7 \times 0.3 = 0.21$.

The probability that he is stopped at the second but not the first is the same.

Note that the sum of probabilities = 1. This is always a good check to do.

Worked example

Venn diagrams

3 A tutor group in a sixth form college comprises 25 students.

The Venn diagram illustrates the numbers in the group studying Mathematics (M), Physics (P) or Chemistry (C).

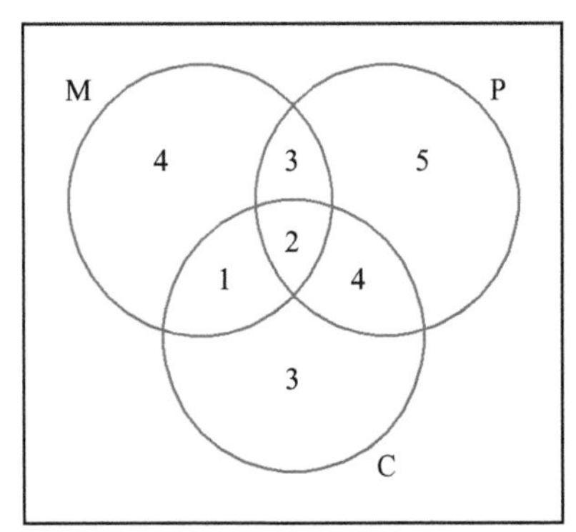

(i) How many study Mathematics?

(ii) How many study Physics but not Chemistry?

(iii) How many of the tutor group do not study any of these subjects?

The three overlapping circles represent the three subjects.

Solution

(i) Inside the 'M' circle there are $4 + 3 + 1 + 2 = 10$, so 10 study Mathematics.

(ii) There are two regions in the P circle that do not overlap with the C circle, so $3 + 5 = 8$ study Physics but not Chemistry.

(iii) All seven regions give a sum of 22, so $25 - 22 = 3$ do not study any of these subjects.

The seven regions represent those who study one, two or three of the subjects; anyone else in the tutor group would be outside these regions.

Exam-style question

TESTED

In one hour on one morning, 45 people go into a coffee shop.
Of these people, 30 buy coffee, 10 buy biscuits and 6 buy sandwiches.
No one buys both biscuits and sandwiches.
Of those who buy coffee, 5 also buy biscuits and 2 buy sandwiches.

(i) Draw a Venn diagram to illustrate this information.
(ii) Determine how many people entered the coffee shop but did not buy any of these three items.

Short answer on page 125

Full worked solution online

CHECKED ANSWERS

10.2 Factorials, permutations and combinations

REVISED

Key facts

1 Factorial notation

$1 \times 2 \times 3 \times \ldots \times n = n!$ This is called 'n factorial'.

$$(m+1) \times (m+2) \times \ldots \times n = \frac{1 \times 2 \times 3 \times \ldots \times m \times (m+1) \times \ldots \times n}{1 \times 2 \times 3 \times \ldots \times m}$$

$$= \frac{n!}{m!}$$

Multiply and divide by all numbers up to and including m.

2 Permutations

A selection of objects in which the order is important is called a permutation.
The number of permutations of r objects from n is denoted by ${}_nP_r$ and given by

$${}_nP_r = \frac{n!}{(n-r)!}$$

The notation ${}_nP_r$ can also be written ${}^nP_r$.

For example, the number of ways that gold, silver and bronze medals can be awarded to the first three runners in a race with seven competitors is $7 \times 6 \times 5$.

$${}_7P_3 = \frac{7!}{(7-3)!} = \frac{7!}{4!} = 210$$

3 Combinations

A selection of objects in which the order is not important is called a combination.
The number of combinations of r objects from n is denoted by ${}_nC_r$ and given by

$${}_nC_r = \frac{n!}{r!(n-r)!}$$

For example, the number of ways that the first three runners can qualify for the next round is $\frac{7 \times 6 \times 5}{1 \times 2 \times 3} = 35$.
There are 3! ways in which each combination will appear in a list of permutations.

The notation ${}_nC_r$ can also be written ${}^nC_r$ or as $\binom{n}{r}$.

Worked example

Permutations

1 From a group of 40 people, the roles of Chairman, Secretary and Treasurer are to be selected.

In how many ways can this be done?

In this example, order matters so it is a permutation.

Solution

The Chairman can be chosen in 40 ways. There are then 39 choices for Secretary and 38 for Treasurer.

$${}_{40}P_3 = \frac{40!}{(40-3)!} = 40 \times 39 \times 38$$

So the number of ways is $40 \times 39 \times 38 = 59280$.

Worked example

Combinations

2 From a group of 40 people in a club, 3 are to be selected to represent the group at a county meeting.

In how many ways can this be done?

In this example order does not matter so it is a combination.

Solution

There are $40 \times 39 \times 38 = 59280$ ways of selecting 3 people in order.

However, in this case order does not matter.

So the 59280 ways will include sets of the same 3 people repeated.

For every 3 people there will be 3! ways.

So the number of selections is

$$\frac{40 \times 39 \times 38}{1 \times 2 \times 3} = \frac{59280}{6} = 9880$$

$${}_{40}C_3 = \frac{40!}{(40-3)!3!} = \frac{59280}{6} = 9880$$

Worked example

Selections and probability

3 A die is tossed 4 times and the scores are written down in a row in order.
(i) How many possible numbers can the four scores make?
(ii) What is the probability that all 4 scores are the same?

Solution

(i) The first score could be any number from 1 to 6, i.e. 6 different numbers.
The second score could also be 1, 2, 3 up to 6.
The same is true for the third and fourth scores.
So the number of numbers possible is $6 \times 6 \times 6 \times 6 = 6^4 = 1296$.

(ii) Of these 1296 numbers one will be 1111, one will be 2222 and so on, i.e. 6 numbers.
So, the probability that one of these numbers appears is $\frac{6}{1296} = \frac{1}{216}$

Exam-style question

TESTED

A group of 6 friends go to the theatre.

(i) How many ways are there for them to sit in a row of seats in the theatre with no restrictions?
(ii) Mia sits in the seat at the left-hand end of the row. How many ways are there for the remaining 5 to seat themselves?
(iii) Additionally, Jane and Julian sit next to each other. Now how many ways are there for the group to sit?

Short answer on page 125

Full worked solution online

CHECKED ANSWERS

Chapter 11 The binomial distribution

About this topic

The binomial expansion is the expression obtained when you multiply out $(a+b)^n$ where n is a positive integer. This expansion is also used in algebra when powers of brackets are required.

An application of the binomial expansion is the binomial distribution in probability and statistics. This is used to model situations in which a number of independent trials are carried out, each with two possible outcomes that can be described as success and failure. The probability of success is conventionally denoted by p, that of failure by q and the number of trials by n. So the binomial distribution is written as $(p+q)^n$. Each term in the expansion gives the probability of a particular number of successes.

Before you start, remember ...

- how to expand brackets
- indices.

11.1 The binomial expansion

REVISED

Key facts

1 **The binomial expansion**
When $(a+b)^n$ is expanded the first term is a^n, the last is b^n and the middle terms take the form ${}_nC_r\, a^r b^{n-r}$. The powers of a and b sum to n and ${}_nC_r$ is the coefficient.

2 **Binomial coefficients**
There are several ways of finding binomial coefficients

- Pascal's triangle

$$\begin{array}{ccccccccccc} & & & & & 1 & & & & & \\ & & & & 1 & & 1 & & & & \\ & & & 1 & & 2 & & 1 & & & \\ & & 1 & & 3 & & 3 & & 1 & & \\ & 1 & & 4 & & 6 & & 4 & & 1 & \\ 1 & & 5 & & 10 & & 10 & & 5 & & 1 \end{array}$$

etc.

- Use of the formula

$${}_nC_r = \frac{n!}{r!(n-r)!}.$$

- The first few terms
For $r = 1, 2, 3\ldots$ the coefficients are

$$1, \frac{n}{1}, \frac{n(n-1)}{1.2}, \frac{n(n-1)(n-2)}{1.2.3}, \ldots$$

The general coefficient is also written as ${}^nC_r$ or as $\binom{n}{r}$.

This is valuable when many or all of the terms are required and n is not too large.

Each coefficient can be generated by adding the two above.

Remember that $n! = n \times (n-1) \times (n-2) \times \ldots \times 3 \times 2 \times 1$.

Note: If n is large and only the first few terms are required, then this method is much more efficient than Pascal's triangle because in order to find a few coefficients, the whole of Pascal's triangle must be created.

Worked examples

The binomial expansion

1 Expand $(2+x)^5$.

Solution

The coefficients are obtained from the row of Pascal's triangle that starts 1 5...

They are 1, 5, 10, 10, 5, 1.

$$(2+x)^5 = 2^5 + 5\times2^4\times x + 10\times2^3\times x^2 + 10\times2^2\times x^3 + 5\times2^1\times x^4 + x^5$$
$$= 32 + 160x + 80x^2 + 40x^3 + 10x^4 + x^5$$

See Pascal's triangle shown earlier.

Don't forget to get the powers of the '2' correct!

Remember that the powers of each term add up to n, in this case 5.

2 Find the first four terms of the expansion $(1-2x)^8$.

Solution

$$(1-2x)^8 = 1 + \frac{8}{1}(-2x) + \frac{8.7}{1.2}(-2x)^2 + \frac{8.7.6}{1.2.3}(-2x)^3 + $$
$$= 1 - 8\times2x + 28\times4x^2 - 56\times8x^3 + \ldots$$
$$= 1 - 16x + 112x^2 - 448x^3$$

The coefficients can be obtained from Pascal's triangle but can also be calculated directly.

Take care with the negative sign.

Exam-style question

TESTED

(i) Expand $(1-x)^{12}$ in ascending powers of x up to the term in x^3, and simplify your answer.

(ii) Find the coefficient of the constant term in the expansion of $\left(2x-\frac{3}{x}\right)^8$.

Short answer on page 125

Full worked solution online

CHECKED ANSWERS

11.2 The binomial distribution

REVISED

Key facts

1 **The binomial distribution**

The binomial distribution is used to calculate the probabilities of different numbers of successes in situations where a number of trials are carried out, each with two possible outcomes that can be described as success and failure. The probability of success is conventionally denoted by p, that of failure by $q = 1 - p$ and the number of trials by n.

The probability of r successes is ${}_nC_r p^r q^{n-r}$.

Note in this term
(i) the sum of powers is n,
(ii) the inclusion of a coefficient.

2 **Conditions for the binomial distribution**

- The binomial distribution involves a number of modelling assumptions. It is a valid model for calculating probabilities of the outcomes of n trials if:
- There are two outcomes, p and q where $p + q = 1$.
- The probability of success, p, is constant for all trials.
- The probability of success is independent of the outcome of any previous trial.

Worked examples

The binomial distribution

1 The probability of obtaining a head when a coin is tossed is $\frac{1}{2}$.

(i) If the coin is tossed 10 times, what is the probability that all 10 are heads?

(ii) What is the probability of obtaining an equal number of heads and tails?

Solution

In this case the conditions for the binomial distribution are met.

(i) The probability of obtaining 10 heads is $\left(\frac{1}{2}\right)^{10}$

This is the first term ${}_{10}C_{10}p^{10}q^0 = p^{10}$.

(ii) The 'middle' term is ${}_nC_r p^r q^{n-r} = {}_{10}C_5 p^5 q^5 = 252\left(\frac{1}{2}\right)^5\left(\frac{1}{2}\right)^5$

$$= \frac{252}{2^{10}} = 0.246$$

2 A normal die is thrown 3 times. What is the probability of obtaining

(i) one six

(ii) at least one six?

Solution

(i) the conditions for the binomial distribution have been met.

$$p = \frac{1}{6}, q = \frac{5}{6}$$

$$P(\text{one six}) = {}_3C_1 pq^2 = 3\left(\frac{1}{6}\right)\left(\frac{5}{6}\right)^2$$

$$= 0.347$$

(ii) P(at least one six) = 1 − P(0 sixes)

$$= 1 - \left(\frac{1}{6}\right)^3 = 1 - 0.0046 = 0.995$$

Either there are no sixes or there is at least one.

It is usually more efficient to use the fact that these two probabilities cover all possibilities and so add to 1.

3 A normal die is thrown 5 times. What is the probability of obtaining exactly 3 sixes?

Solution

$$P(\text{six}) = \frac{1}{6}$$

$$\Rightarrow P(3 \text{ sixes in } 5 \text{ throws}) = {}_5C_3\left(\frac{1}{6}\right)^3\left(\frac{5}{6}\right)^2$$

$$= \frac{5!}{3!2!} \times \frac{1}{216} \times \frac{25}{36} = 0.0322$$

$P(\text{six}) = \frac{1}{6}$, $P(\text{not a six}) = \frac{5}{6}$.

Note that the powers sum to 5 and the coefficient found as here or from Pascal's triangle (10).

Exam-style question

TESTED

Glass marbles are produced in two colours, red and green, in the proportion 7 : 3 respectively. From a large stock of marbles, 5 are taken at random.

(a) Using the binomial distribution, find the probability that
 (i) all 5 are red,
 (ii) exactly 2 are red.

(b) Explain why the binomial distribution is appropriate in this case.

Short answer on page 125

Full worked solution online

CHECKED ANSWERS

Review questions (Chapters 10–11)

1 A fair die is rolled twice. The 'score' is double the first number plus the second number.

(i) Draw a sample space to illustrate the possible outcomes. **[2]**

(ii) Find the probability that the score is 12. **[2]**

2 Ahmed has a bag containing 10 balls, 5 of which are blue and the other 5 are red. He draws a ball at random followed by a second ball at random.

(i) Draw a probability tree to show the possible outcomes. **[2]**

(ii) What is the probability that the first ball is red and the second is blue? **[2]**

(iii) What is the probability that the two balls are one of each colour? **[2]**

3 (i) How many ways are there of choosing at random a committee of 5 people from 10 people? **[2]**

(ii) If, first, the chairman is chosen at random before the others are chosen at random, how many ways are there of forming the committee of five? **[4]**

4 Seven members of a team are to have their photograph taken while sitting on a line of seven chairs.

(i) In how many ways can they be sat if the captain sits on the middle chair? **[2]**

(ii) Additionally, Pedro and Quentin, being the tallest, are to sit on the end seats.

How many ways can the team now sit? **[2]**

5 Expand $(4x-1)^5$ simplifying all the terms. **[3]**

6 The first three terms in the expansion of $(1+ax)^n$ are $1+3x+\frac{15}{4}x^2$. **[5]**

Find the values of *a* and *n*.

7 (i) Write down the first four terms in the expansion of $\left(1+\frac{x}{2}\right)^8$. **[3]**

(ii) Expand $\left(1-\frac{2}{x}\right)^2$. **[2]**

(iii) Hence find the constant term in the expansion of $\left(1-\frac{2}{x}\right)^2\left(1+\frac{x}{2}\right)^8$. **[4]**

8 Tracey tosses a fair coin 4 times.
What is the probability that

(i) she obtains exactly 2 heads, **[3]**

(ii) at least one tail is obtained? **[3]**

9 The workforce of a large company is made up of males and females in the ratio 9:11. One third of the male employees work part-time and one half of the females work part-time.
Eight employees are chosen at random.
Find the probability that

(i) all are males, **[2]**

(ii) exactly five are females, **[4]**

(iii) at least two work part-time. **[5]**

Short answers on page 126

Full worked solutions online

CHECKED ANSWERS ☐

Section 5 Powers and iteration

Target your revision (Chapters 12–13)

Try answering each question below. If you get stuck, follow the page reference underneath to revise that topic.

1 **Curves of exponential functions**
Identify which of the following equations fit the curves shown.
A $y = 2 \times 3^x$
B $y = 2 + 3^x$
C $y = 3 \times 2^{-x}$

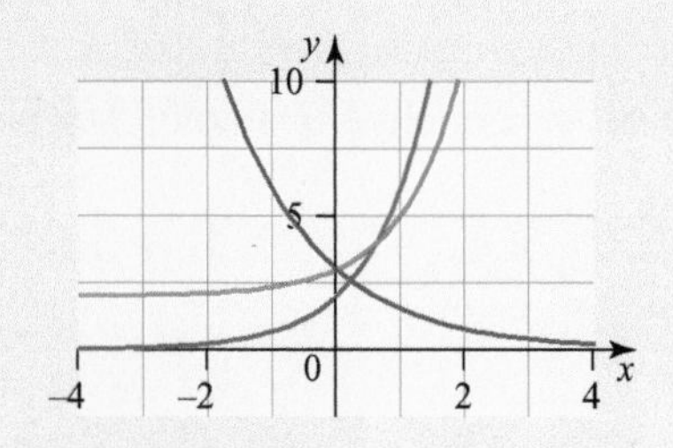

(see page 73)

2 **Exponential functions**
When taking a particular medicine, the amount, *m* milligrams, left in the bloodstream after *t* hours, can be modelled by the equation $m = 20 \times 1.7^{-0.8t}$.
(i) What is the initial dose?
(ii) How much medicine is still in the bloodstream after 6 hours, correct to 3 significant figures?

(see page 73)

3 **Logarithms**
Write $3\log 2 - \frac{1}{2}\log 14$ as a single logarithm.

(see page 76)

4 **Reduction of $y = ka^x$ to linear form**
(i) By taking logarithms, show that the data in the table below can be well modelled by the equation $y = ka^x$.
(ii) Plot a graph and draw a line of best fit.
(iii) Use your line to find the values of *k* and *a*.

x	1	1.5	2	2.5	3
y	0.375	0.650	1.125	1.950	3.375

(see page 77)

5 **Reduction of $y = kx^n$ to linear form**
By taking logarithms, show that the data in the table below can be well modelled by the equation $y = kx^n$ and find the values of *k* and *n*.

x	1	1.5	2	2.5	3
y	0.1	0.5	1.6	3.9	8.1

(see page 77)

6 **Simple exponential equations**
Solve the equation $2^x = 7$

(see page 80)

7 **More complicated exponential equations**
Solve the equation $2^{x+1} = 3^{2-3x}$

(see page 80)

8 **Locating a root of an equation by change of sign**
Use decimal search to find the root of the equation $x^3 + x - 3 = 0$, correct to 2 decimal places.

(see page 82)

9 **Interval bisection**
Use the bisection method to find the root of the equation $x^3 - 2x - 2 = 0$, correct to 2 decimal places.

(see page 82)

10 **Iterative sequences**
You are given the equation $f(x) = x^3 - x - 3$.
(i) Show by a sketch and change of sign that there is a root in the interval [1, 2].
(ii) Show that the equation can be rewritten as $x = \sqrt[3]{x+3}$.
(iii) Using the iterative formula $x_{r+1} = \sqrt[3]{x_r + 3}$ with $x_0 = 1$, find the root correct to 4 decimal places.

(see page 85)

11 Gradients of tangents – central estimate

You are given that the points A (1, 0.42), B (2, 1.04) and C (3, 2.60) lie on the curve $y = \frac{1}{6} \times 2.5^x$. Coordinates have been given to 2 decimal places.

(i) Sketch the curve from $x = 0$ to $x = 3$ and add the points A, B and C.

(ii) Find the gradient of the chord AC and explain why this is an estimate of the gradient of the tangent at B.

(see page 87)

12 Gradients of tangents – one-sided estimate

You are given that the points A (1, 0.42), B (2, 1.04) and C (3, 2.60) lie on the curve $y = \frac{1}{6} \times 2.5^x$

(i) Find the gradients of the chords AB and BC and explain how these can be used to find an estimate for the gradient of the tangent at B.

(ii) Comment on the accuracy and explain how the accuracy could be improved.

(see page 87)

13 Area under a curve by rectangles.

The curve $y = 8 \times 2^{-x}$ is shown below.

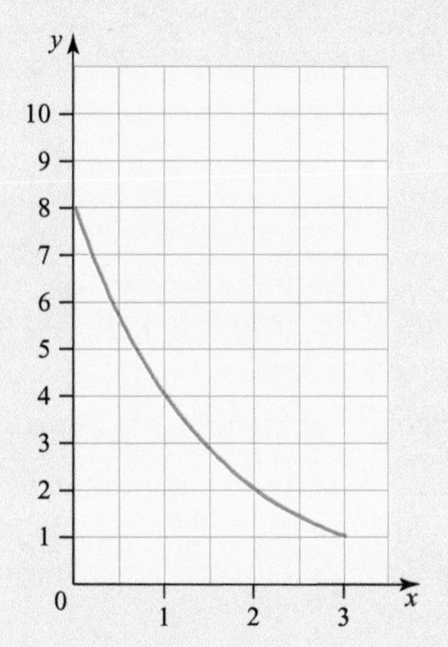

Find an approximation to the area between the curve, the x-axis, the y-axis and the line $x = 3$ using

(i) three rectangles beneath the curve,

(ii) 3 rectangles above the curve.

(iii) Hence give an estimate for the area.

(see page 88)

14 Area under a curve by trapezia

Find an approximation to the area between the curve $y = 8 \times 2^{-x}$, the x-axis, the y-axis and the line $x = 3$ using 3 trapezia.

(see page 88)

CHECKED ANSWERS

Short answers on pages 126–27

Full worked solutions online

Chapter 12 Exponentials and logarithms

About this topic

An exponential function in x has the form ka^x or ka^{-x}.

Many real-life examples of growth (e.g. population growth) or decay (e.g. radioactive decay) can be modelled by exponential functions.

The inverse of an exponential function is a logarithm.

Several well-known measures use a logarithmic scale, for example the Richter scale for the strength of earthquakes and measurements of the loudness of sound in decibels. In this chapter you will see how logarithms are used in calculations.

Before you start, remember ...

- indices.

12.1 Exponential functions

REVISED

Key facts

1. **The exponential function**
   The function $y = ka^x$ is called an exponential function.
   The variable x is in the power, or the exponent.
2. **Properties of exponential functions of the form $y = ka^x$ ($a > 1$).**
   - They pass through the point (0, k)
   - The gradient is positive for all x
   - The negative x-axis is an asymptote
   - $y > 0$ for positive k and $y < 0$ for negative k.
3. **Properties of exponential functions of the form $y = ka^{-x}$ ($a > 1$)**
   - They pass through the point (0, k)
   - The gradient is negative for all x
   - The positive x-axis is an asymptote
   - $y > 0$ for positive k and $y < 0$ for negative k.

The graph of $y = ka^x$.

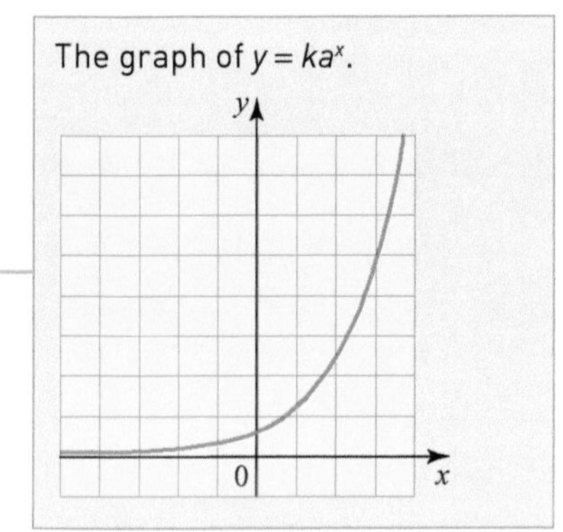

The graph of $y = ka^{-x}$.

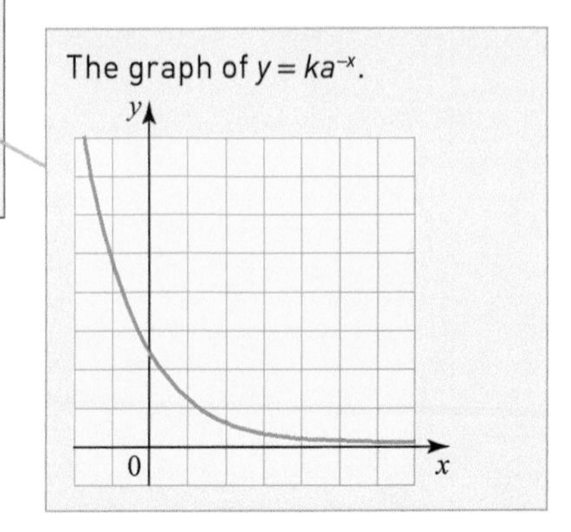

Worked examples

The graph of exponential functions

1 Sketch the graphs of the following.

(i) $y = 2 \times 3^x$

(ii) $y = 1 + 2 \times 3^x$

Solution

(i)

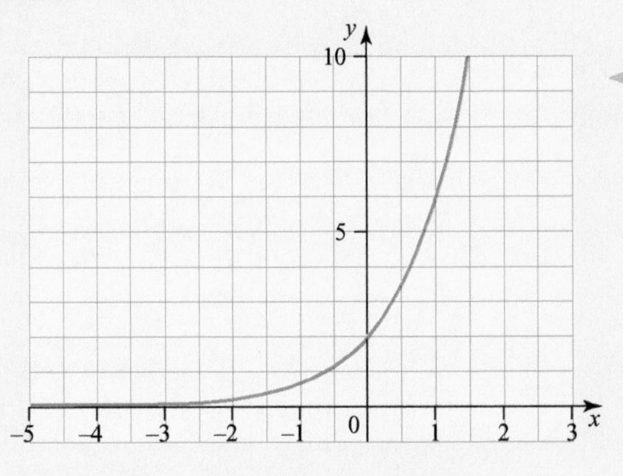

This graph has the same shape as the one above with the same properties except that it passes through (0, 2).

(ii)

This graph is parallel to the one in (i) but is translated one unit in the positive y direction. Note that because the distance between the curves is constant the graphs are parallel, though they do not look it!

2 (i) Draw $y = 2 \times 3^{-x}$

(ii) Draw $y = 2 \times \left(\frac{1}{3}\right)^x$

(iii) Explain why your two graphs are the same.

Solution

(i) and (ii): the graph is as shown.

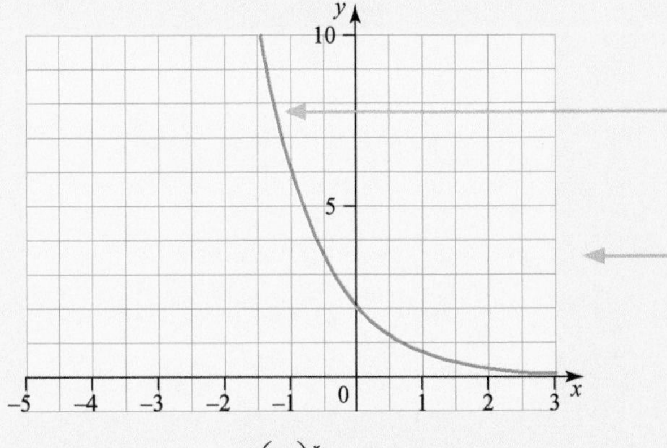

This curve is an example of exponential decay. The graph has negative gradient throughout, it passes through (0, 2) and the positive x-axis is an asymptote.

Note that the graph $y = ka^x$ for $k > 0$ will represent decay when $0 < a < 1$.

(iii) $(3)^{-x} = (3^{-1})^x = \left(\frac{1}{3}\right)^x$ so the functions are the same.

Worked example

Exponential growth

3 Two rabbits are put in an enclosure. The population, P, after t months is modelled by the equation $P = ka^{t/4}$.

After 4 months there are 6 rabbits.

Find the size of the population after 12 months.

Solution

Substitute initial conditions.

$P = ka^{t/4}$

Given $P = 2$ when $t = 0$

$P = ka^0 = k$

$\Rightarrow k = 2$

Given $P = 6$ when $t = 4$

$\Rightarrow 6 = 2a \qquad \Rightarrow a = 3$

$\Rightarrow P = 2 \times 3^t$

When $t = 12$, $P = 2 \times 3^{12/4} = 2 \times 3^3 = 54$

Rabbits can become sexually active after about 4 months!

Remember that $a^0 = 1$.

Note that in cases like this, k is the 'initial' value of P. So this is often written as $P = P_0a^t$.

You need two 'boundary' conditions to find the two constants k and a. Substitute the second condition to find a.

Note that population growth modelled as an exponential function soon causes problems – how many rabbits would there be after 2 years?

Exam-style question

TESTED

The value of a car, bought as new, depreciates exponentially. Paula buys the car for £16 000. She discovers that it is worth only £12 000 after one year. What will be the value of her car after 5 years?

Short answer on page 127

Full worked solution online

CHECKED ANSWERS

12.2 Logarithms

REVISED

Key facts

1 **Definition**

The logarithmic function is the inverse of the exponential function.

If $y = a^x$ then $x = \log_a y$ where a is the base of the logarithm.

2 **Logarithms on your calculator**

There are two logarithm keys on most calculators:

- log gives a logarithm to the base 10
- ln gives a natural logarithm; the base is the number e or 2.718... .

Some calculators have a third key which allows you to find a logarithm to any base.

Natural logarithms are beyond the scope of this course.

3 **Graphs of logarithms**

Since logarithms are in the inverse of exponentials, the curve $y = \log_a x$ is a reflection of $y = a^x$ in the line $y = x$.

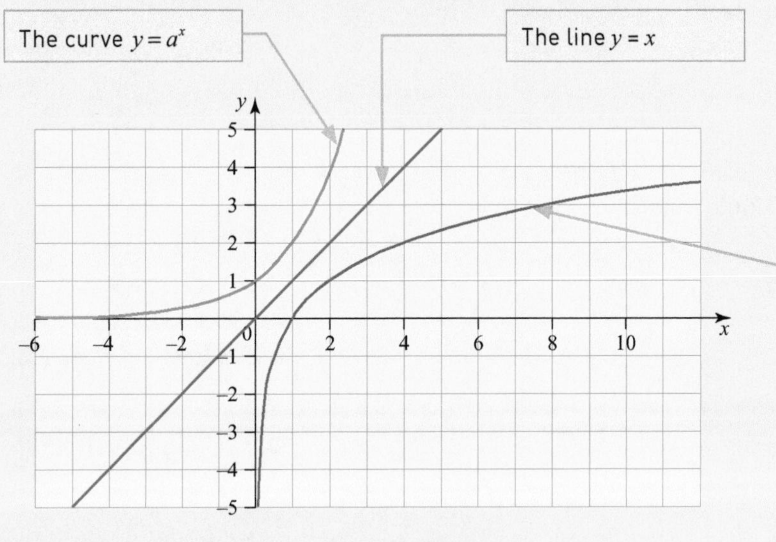

The curve $y = \log_a x$

4 **Properties of the graph of the log function**

For the graph of $y = k \log_a x$ where $a > 1$

- all curves pass through the point (1, 0)
- the curves only exist for positive values of x
- all curves have a positive gradient
- the negative y-axis is an asymptote.

5 **Laws of logarithms**

$$\log xy = \log x + \log y$$

$$\log\left(\frac{x}{y}\right) = \log x - \log y$$

$$\log x^n = n \log x$$

$$\log 1 = 0$$

$$\log_a a = 1$$

Worked example

Graphs of log functions

1 (i) Sketch on the same grid the graphs of $y = \log x$ and $y = \log 2x$.
(ii) What is the relationship between the two curves?

Since the base of the logarithm is not stated it can be assumed to be 10.

Solution

(i)

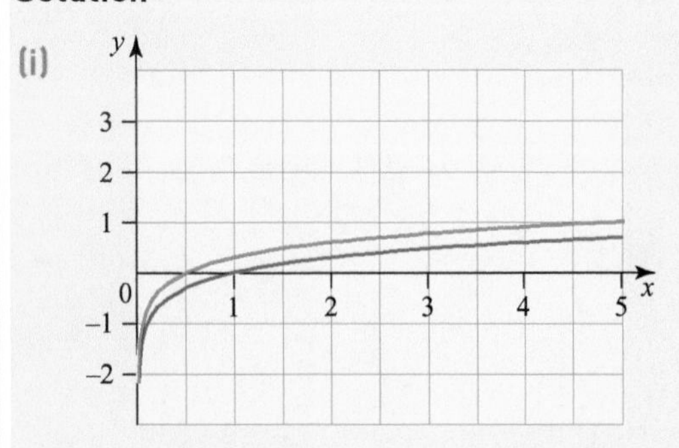

(ii) $y = \log 2x = \log 2 + \log x$, so the second graph is parallel to the first but is translated $\log 2$ units in the positive y direction.

Worked example

Laws of logarithms

2 Write each of the following as a single logarithm.
(i) $\log 9 - 3\log 2$
(ii) $\frac{1}{3}\log 27$

Solution

(i) $\log 9 - 3\log 2 = \log 9 - \log 8$

First use the rule $n\log a = \log a^n$

$= \log\left(\frac{9}{8}\right)$

Then $\log a - \log b = \log\left(\frac{a}{b}\right)$

(ii) $\frac{1}{3}\log 27 = \log(27)^{\frac{1}{3}} = \log 3$

$(27)^{\frac{1}{3}} = \sqrt[3]{27} = 3$

Exam-style question

(i) Write $2\log_{10} x - \log_{10} 6$ as a single logarithm.
(ii) Hence solve the equation $2\log_{10} x - \log_{10} 6 = 1.2$.

Short answer on page 127

Full worked solution online

CHECKED ANSWERS

12.3 Reduction to linear form

REVISED

Key facts

1 **Transforming curves into straight lines**

If you have data which you think satisfy one of the relationships $y = ka^x$ or $y = kx^n$, you can test this by plotting a suitable logarithmic graph and seeing if you get a straight line.

If you do, you can use your line to estimate the values of the constants in the relationship.

2 **Relationships of the form $y = ka^x$**

Take logs of both sides $y = ka^x \Rightarrow \log y = \log k + x \log a$.

So, a graph of $\log y$ against x will be a straight line with gradient $\log a$, and intercept $\log k$ on the vertical axis.

3 **Relationships of the form $y = kx^n$**

Take logs of both sides.

$$y = kx^n \Rightarrow \log y = \log k + n \log x.$$

So a graph of $\log y$ against $\log x$ will be a straight line with gradient n, and intercept $\log k$ on the vertical axis.

Worked example

Relationship of the form $y = ka^x$

1 The data in the table below are thought to be well modelled by the relationship $y = ka^x$.

By drawing a suitable graph, find the values of k and a and so write down the relationship.

x	0	1	2	3
y	0.25	0.75	2.25	6.75

Solution

$y = ka^x \Rightarrow \log y = \log k + x \log a$

So, plot $\log y$ against x.

x	0	1	2	3
y	0.25	0.75	2.25	6.75
$\log y$	−0.60	−0.12	0.35	0.83

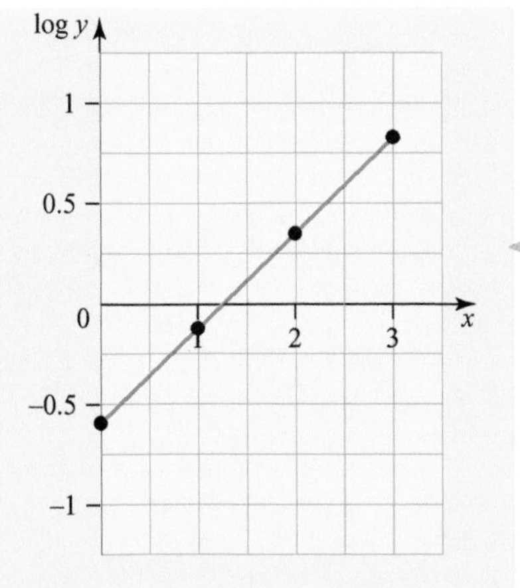

Draw a line of best fit. The intercept on the y axis is $\log k$ and the gradient is $\log a$.

This solution is worked in base 10, but any base would work.

The points all lie on a straight line meaning that the model proposed is a good model.

Intercept on the vertical axis

$-0.6 = \log k$

$k = 10^{-0.6} = 0.251...$

i.e. $k = 0.25$ to 2 significant figures

Gradient

$$\frac{0.83 - -0.6}{3 - 0} = \frac{1.46}{3} = 0.486...$$

$\Rightarrow \log a = 0.486...$

$\Rightarrow \quad a = 3.066...$

$\Rightarrow a = 3.1$ to 2 significant figures

So, the relationship is $y = 0.25 \times 3.1^x$

Worked example

Relationship of the form $y = kx^n$

2 The data in the table below are thought to be well modelled by the relationship $y = kx^n$.

x	1	2	3	4
y	0.5	4	13.5	32

By drawing a suitable graph, find the relationship.

Solution

$y = kx^n$

Take logs of both sides.

$\log y = \log k + n \log x$

So plotting $\log y$ against $\log x$ should give a straight line with intercept $\log k$ on the vertical axis and gradient n.

x	1	2	3	4
y	0.5	4	13.5	32
$\log x$	0	0.3	0.5	0.6
$\log y$	−0.3	0.6	1.1	1.5

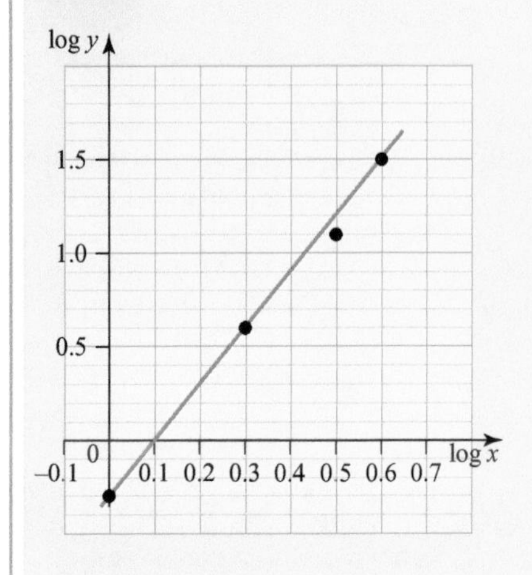

Draw a line of best fit. The intercept on the y-axis is $\log k$ and the gradient is n.

Intercept

The line passes through (0, −0.3)

$\Rightarrow \log k = -0.3$

$\Rightarrow \quad k = 10^{-0.3}$

$\Rightarrow \quad k = 0.501...$

Gradient

$\frac{1.5 - (-.3)}{0.6 - 0} = \frac{1.8}{0.6} = 3$

So $k = 0.5$ and $n = 3$.

The relationship is $y = 0.5x^3$

Exam-style question

TESTED

A new butcher in a small town records the numbers of customers he serves each week. He thinks that if the numbers can be modelled as an exponential increase this will be an indication that the residents of the town like his meat.

The data are in the table.

Week number	1	2	3	4	5	6
Number of customers	120	144	172	208	248	300

(i) Test whether the data satisfy the relationship $y = ka^x$.
(ii) Estimate the values of a and k.
(iii) Explain why the model cannot continue for very long.

Short answer on page 127

Full worked solution online

CHECKED ANSWERS

12.4 Solving exponential equations

REVISED

Key fact

1 **Using logarithms to solve exponential equations**
Equations of the form $a^x = b$ can be solved by taking logs of both sides.

$$a^x = b \Rightarrow x \log a = \log b$$

$$\Rightarrow x = \frac{\log b}{\log a}$$

In these cases, any base can be used but your calculator will give you logs to the base 10, so this is the most obvious and convenient.

Worked examples

Using logs to solve equations

1 Solve the equation $5^x = 15$

Solution

Take logs to base 10

$$5^x = 15 \Rightarrow x \log 5 = \log 15$$

$$\Rightarrow x = \frac{\log 15}{\log 5} = 1.683$$

Check: you know that $5^1 = 5$ and $5^2 = 25$ so $1 < x < 2$, but it is **not** halfway.

2 Solve the equation $2^{1-x} = 7$.

Solution

$$2^{1-x} = 7 \quad \Rightarrow \log 2^{1-x} = \log 7$$

Take logs of both sides.

$$\Rightarrow \quad (1-x)\log 2 = \log 7$$

Use the law $n \log a = \log a^n$.

$$\Rightarrow \quad x = 1 - \frac{\log 7}{\log 2} = -1.81$$

Express x as the subject.

Worked example

Combining log laws with equations

Solve the equation $2^{3x-1} = 3^{x+1}$.

Solution

Take logs of both sides

$2^{3x-1} = 3^{x+1}$

$\Rightarrow \quad (3x-1)\log 2 = (x+1)\log 3$

$\Rightarrow 3x\log 2 - x\log 3 = \log 3 + \log 2$

$\Rightarrow x(3\log 2 - \log 3) = \log 3 + \log 2$

$\Rightarrow \quad x(\log 8 - \log 3) = \log 3 + \log 2$

$\Rightarrow \quad x\log\left(\frac{8}{3}\right) = \log 6$

$\Rightarrow \quad x = \frac{\log 6}{\log\left(\frac{8}{3}\right)} = 1.83$

Take logs of both sides.

Use the law $n\log a = \log a^n$.

Collect like terms.

Use the law $\log a + \log b = \log ab$ and $\log a - \log b = \log\left(\frac{a}{b}\right)$.

Warning: A fraction involving a logarithm on the top and bottom cannot be found using the log laws – they are two numbers which need to be divided.

Exam-style question

TESTED

It is thought that the temperature, T°C , of a cup of tea t minutes after it was made with boiling water can be modelled by the equation $T - 25 = 75 \times a^{-\frac{t}{4}}$ where a is an integer.

(i) Show, by taking logs, how this model can be tested.
The data for this cup are given in the following table.

Time (minutes)	2	4	6	8	10
Temperature (°Celsius)	68.3	50	39.4	33.3	29.8

(ii) By plotting a straight line, show that the data support the model and estimate the value of a.

(iii) What is the long-term temperature of the cup of water?

Short answer on page 127

Full worked solution online

CHECKED ANSWERS

Chapter 13 Numerical methods

About this topic

As you have seen, some equations can be solved algebraically to give the roots exactly. However, there are many equations for which no algebraic solution exists. In those cases you need to use numerical methods. A numerical method does not give an exact answer but it can be as accurate as you choose to make it.

Numerical methods can also be used for other mathematical tasks, like finding the gradient at a point on a curve and finding the area under a curve. Both of these underpin the development of calculus, which you will meet in Chapters 14 and 15.

Numerical methods are often carried out on computers and so have become increasingly important in recent years.

Before you start, remember ...

- how to solve an equation
- how to find a gradient
- how to find the area of a trapezium.

13.1 Sign change methods for solving an equation

REVISED

Key facts

1 **Locating the root of an equation from a graph**
If the graph of the function $y = f(x)$ crosses the x-axis at the point $x = \alpha$, then this is a root of the equation $f(x) = 0$.
The answer is only as accurate as the graph.

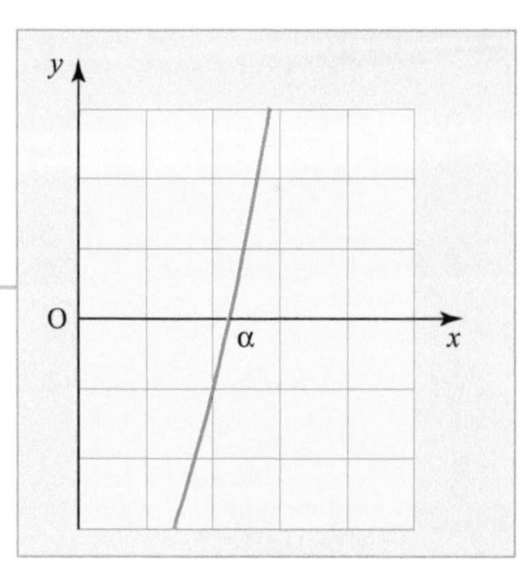

2 **Change of sign**
If the curve $y = f(x)$ crosses the axis at $x = \alpha$, then on one side of the root the value of the function will be positive and on the other it will be negative. By simple calculation the root can be 'trapped' between two values of x, giving the answer as accurately as you wish.

> For example, to find the root of $f(x)=1+\sqrt{x}-x^2=0$,
> $f(1) = 1 > 0$, $f(2) = -1.586 < 0$ so the graph of $f(x)=1+\sqrt{x}-x^2=0$ crosses the axis in the range [1, 2]. There is therefore a root within that range.
> This root can be given as $x = 1.5 \pm 0.5$.
> By decreasing the range you can obtain a more accurate answer.

3 **Decimal search**
This method involves finding a range within which the root lies by change of sign, then decreasing the steps as many times as you need to obtain the required accuracy.

For example, if an integer search finds a root in the interval [1, 2] then start again with $x = 1$ and increase in steps of 0.1 until you find a change of sign.

4 **Interval bisection**
In this method the range is decreased by halving the interval.
For example, if $f(1) > 0$ and $f(2) < 0$ then you know the root is in the interval [1, 2].
So evaluate $f(1.5)$ to see if the root lies in the interval [1, 1.5] or [1.5, 2].

The calculations are a little more complicated but there are two advantages to this method:
(i) convergence can be quicker,
(ii) you know how many iterations you need to ensure an answer to the accuracy required.

Worked example

Decimal search

1 Use decimal search to find the root of the equation $1+\sqrt{x}-x^2=0$, correct to 4 significant figures.

Solution

From above, $1 < x < 2$

From the lower limit of the range, increase the value of x in steps of 0.1 until you find a change of sign.

x	f(x)
1	1
1.1	0.838809
1.2	0.655445
1.3	0.450175
1.4	0.223216
1.5	−0.02526

There is a change of sign between $x = 1.4$ and $x = 1.5$.

You can now say that the root is $x = 1.45 \pm 0.05$.

x	f(x)
1.4	0.223216
1.41	0.199334
1.42	0.175238
1.43	0.150926
1.44	0.1264
1.45	0.101659
1.46	0.076705
1.47	0.051536
1.48	0.026153
1.49	0.000556
1.5	−0.02526

There is a change of sign between $x = 1.49$ and $x = 1.50$.

You can now say that the root is $x = 1.495 \pm 0.005$

This is correct to 4 significant figures.

Note that the root is given to the required accuracy and the error bounds are stated.

Worked example

Interval bisection

2 Use interval bisection to find the root of the equation $1+3\sqrt{x}-x^2=0$, correct to 3 significant figures.

Solution

x	f(x)
2	1.242641
3	−2.80385

The root lies in the interval [2, 3] so can be stated to be $x = 2.5 \pm 0.5$.

x	f(x)
2	1.242641
2.5	−0.50658
3	−2.80385

$f(2.5) < 0$ so the root lies in the interval [2,2.5] so can be stated to be $x = 2.25 \pm 0.25$.

x	f(x)
2	1.242641
2.25	0.4375
2.5	−0.50658

f(2.25) > 0 so the root lies in the interval [2.25, 2.5] so can be stated to be $x = 2.375 \pm 0.125$.

x	f(x)
2.25	0.4375
2.375	−0.01731
2.5	−0.50658

f(2.375) < 0 so the root lies in the interval [2.25, 2.375] so can be stated to be $x = 2.3125 \pm 0.0625$.

x	f(x)
2.25	0.4375
2.3125	0.214416
2.375	−0.01731

f(2.3125) > 0 so the root lies in the interval [2.3125, 2.375] so can be stated to be $x = 2.34375 \pm 0.0231255$.

x	f(x)
2.34375	0.099629
2.375	−0.01731

f(2.34375) > 0 so the root lies in the interval [2.34375, 2.375] so can be stated to be $x = 2.359375 \pm 0.015625$.

2.34375	0.099629
2.359375	0.041427
2.375	−0.01731

f(2.359375) > 0 so the root lies in the interval [2.359375, 2.375] so can be stated to be $x = 2.3671875 \pm 0.0078125$.

2.3671875	0.012123
2.37109375	−0.00258
2.375	−0.01731

f(2.37109375) < 0 so the root lies in the interval [2.3671875, 2.37109375] so can be stated to be $x = 2.36914065 \pm 0.00390625$.

So to three significant figures, the root is 2.37.

Exam-style question

TESTED

The graph shows a curve with equation $x^3 + x - 4 = 0$.

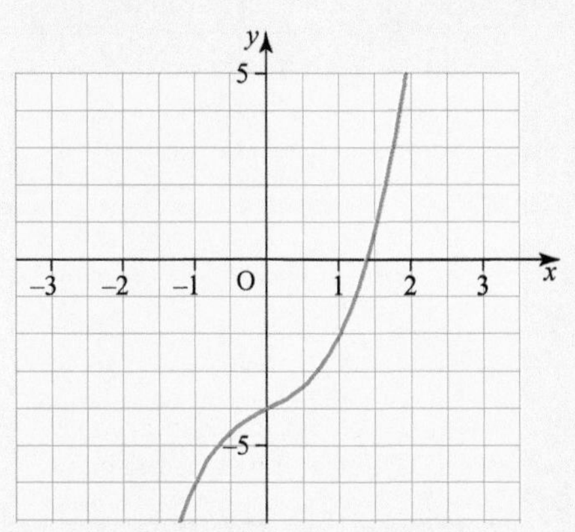

(i) Verify that the equation has a root between $x = 1$ and $x = 2$.
(ii) Using the interval bisection method find this root correct to 2 significant figures.

Short answer on page 127

Full worked solution online

CHECKED ANSWERS

13.2 Iterative methods

REVISED

Key facts

1 **Definition**
In an iterative sequence the first term is given or chosen. Each subsequent term is formed by using the previous term as the input to the rule.

2 **Solving equations**
The equation $f(x) = 0$ is rearranged into the form $x = g(x)$ and the rule for the iteration is $x_{n+1} = g(x_n)$.
The initial value, x_0, is given or chosen.
To be useful an iteration must converge but this does not always happen; some iterations diverge.

The initial value x_0 is used to find x_1, x_1 is then used to find x_2 and so on.

3 **Graphical representation.**
If $x = g(x)$ is a rearrangement of the equation $f(x) = 0$ then the root of the equation is where $y = g(x)$ cuts the line $y = x$.
An iterative sequence to find the root will often take one of two forms:

- A staircase diagram gives an increase (or decrease) to the root as shown in the first diagram.
- A cobweb diagram will occur when the iterative sequence oscillates above and below the root, as shown in the second diagram.

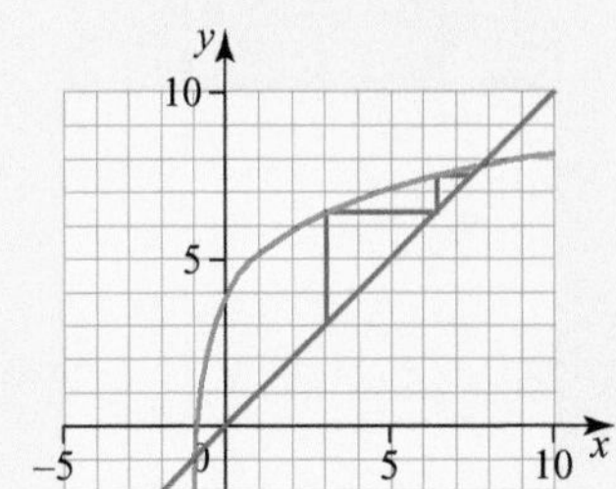

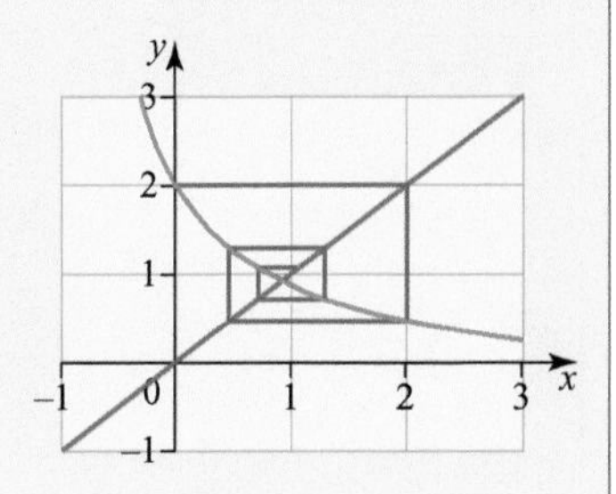

Worked examples

Iterative sequences

1 (i) Show that the equation $x^3 + x - 4 = 0$ can be rearranged to the form $x = \sqrt[3]{(4 - x)}$.
(ii) Using an initial value of $x = 1$ and the iterative sequence $x_{n+1} = \sqrt[3]{(4 - x_n)}$, find the root to 2 significant figures.

Notice the comparison with the decimal search methods above.

Solution

(i) $x^3 + x - 4 = 0$

$\Rightarrow \quad x^3 = 4 - x$

$\Rightarrow \quad x = \sqrt[3]{(4 - x)}$

(ii) $x_{n+1} = \sqrt[3]{(4 - x_n)}$

Taking $x_0 = 1 \Rightarrow x_1 = \sqrt[3]{3} = 1.442$

$\Rightarrow \quad x_2 = \sqrt[3]{(4 - 1.442)} = \sqrt[3]{(2.558)} = 1.368$

$\Rightarrow \quad x_3 = \sqrt[3]{(4 - 1.368)} = \sqrt[3]{(2.632)} = 1.381$

$\Rightarrow \quad x_4 = \sqrt[3]{(4 - 1.381)} = \sqrt[3]{(2.619)} = 1.378$

This shows that, to 2 significant figures, the root is $x = 1.4$.

Substitute the value for x_0 given to find x_1.

Repeat the process to find x_2, x_3, and so on until the value fulfil the accuracy requirement.

This shows that the diagram will be a cobweb diagram which converges.

Check that you can carry out this iterative sequence using the ANS and enter keys.

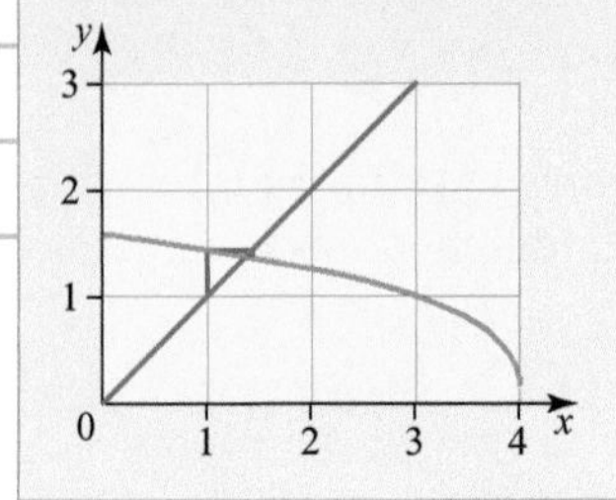

2 (i) Show that the equation $x^3+x-4=0$ can be rearranged to the form $x=\dfrac{4}{x^2+1}$.

(ii) Show that the iterative sequence $x_{n+1}=\dfrac{4}{x_n^2+1}$ starting with an initial value of $x=1$ does not converge to the root.

There are many ways to rearrange an equation. If you have to find one, then it is possible that it does not converge. Finding an iterative sequence can therefore be quite long winded. However, if you find one it is possible that it will converge very fast.

Solution

(i) $x^3+x-4=0$

$\Rightarrow x(x^2+1)=4$

$\Rightarrow \quad x=\dfrac{4}{x^2+1}$

(ii) $x_0=1\Rightarrow x_1=\dfrac{4}{1+1}=2$

$\Rightarrow x_2=\dfrac{4}{4+1}$

$=0.8$

$\Rightarrow x_3=\dfrac{4}{0.64+1}$

$=2.439$

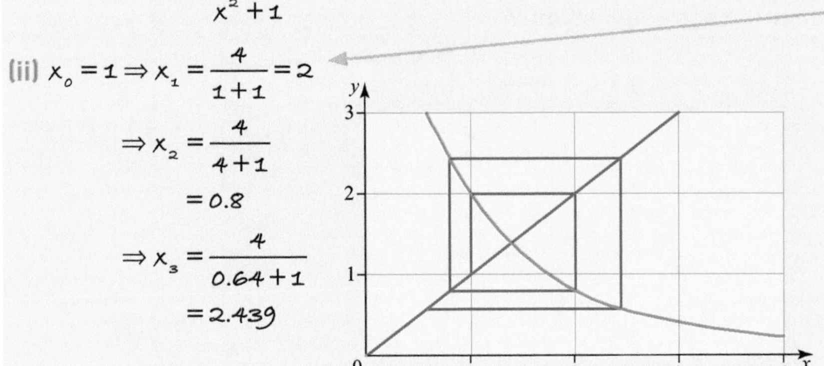

x_0 is below the root and x_1 is above it. So this will give a cobweb diagram.

If the sequence converges then x_2 would be closer to the root than x_0 i.e. greater than 1.

However, it is not and so the sequence will diverge.

3 (i) Show that the cubic equation $x^3-5x-6=0$ can be rearranged to the equation $x=\sqrt[3]{5x+6}$.

(ii) Using an initial value of $x=2$, find the root in the interval [2, 3], correct to 2 decimal places.

Solution

(i) $x^3-5x-6=0$

Make x^3 the subject and then take the cube root.

$\Rightarrow \quad x^3=5x+6$

$\Rightarrow \quad x=\sqrt[3]{(5x+6)}$

(ii)

x_n	x_{n+1}
2	2.519842
2.519842	2.649506
2.649506	2.679939
2.679939	2.686983
2.686983	2.688608

This shows that, to 2 decimal places, the root is $x=2.69$.

Exam-style question

TESTED

(i) Show that the equation $x=10^{x-2}$ can be rearranged to the equation $x=\log x+2$.

(ii) Sketch the curves with equations $y=x$ and $y=\log x+2$.

(iii) Using an initial value of $x=1$ demonstrate on your graph that the iterative procedure $x_{n+1}=\log x_n+2$ converges to the upper root.

(iv) Find the root correct to 3 significant figures.

Short answer on pages 127–28

Full worked solution online

CHECKED ANSWERS

13.3 Gradients of tangents

REVISED

Key facts

1 **Definitions**

(i) A tangent to a curve at a point is the line that touches the curve at that point.

(ii) The gradient of the curve at a point is the gradient of the tangent at that point.

2 **Estimating the gradient from a chord**

The line joining two points on a curve is a chord. The gradient of a chord provides an estimate for the gradient of the curve; the nearer the two points are together the better the estimate.

The gradient of the line through two points (x_1, y_1) and (x_2, y_2) is given by $\frac{y_2 - y_1}{x_2 - x_1}$.

3 **Central estimate**

For a central estimate, the two points are equally spaced either side of the point on the curve.

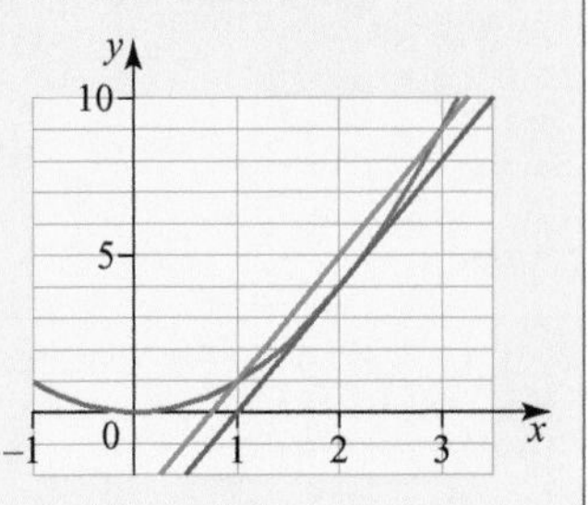

4 **Forward estimate**

For a forward estimate the point on the left is the one where the gradient is to be found.

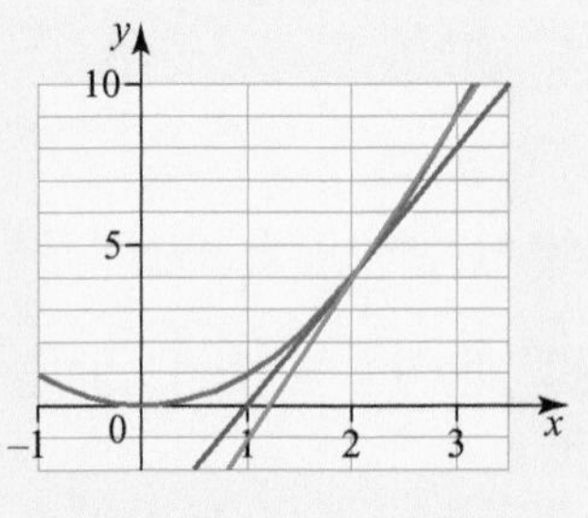

5 **Backward estimate**

For a backward estimate the point on the right is the one where the gradient is to be found.

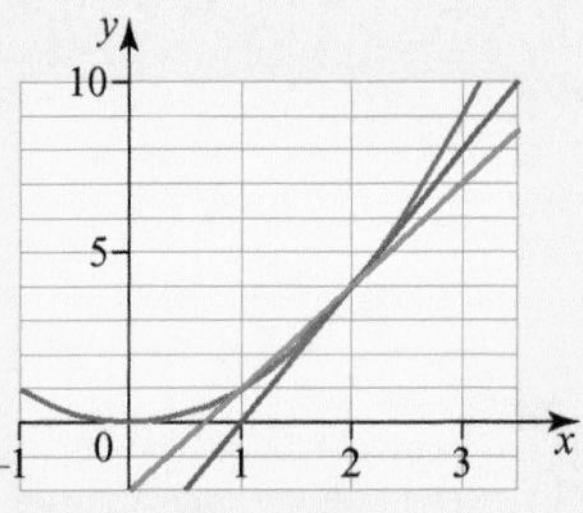

Note: For polynomial curves, calculus can be used to calculate the gradient of a curve at a point. This is described in Chapter 14.

Worked example

Using the three estimates

(i) Find an estimate for the gradient of the curve $y = 2^x$ at the point P (1, 2) using all three methods and the points where $x = 0.9$ and $x = 1.1$.

(ii) Comment on the accuracy of your estimates.

Solution

(i) The backward estimate uses a point to the left of P (1,2).
Take this point to be (0.9, 1.866)

$$\text{Gradient} = \frac{1.866 - 2}{0.9 - 1} = 1.34$$

The forward estimate uses a point to the right of P.
Take this point to be (1.1, 2.144)

$$\text{Gradient} = \frac{2.144 - 2}{1.1 - 1} = 1.435$$

The central estimate uses two points either side of P.
Take these points to be (0.9, 1.866) and (1.1, 2.144).

$$\text{Gradient} = \frac{2.144 - 1.866}{1.1 - 0.9} = 1.39$$

(ii) The backward estimate is too small. The forward estimate is too large and the mean of the two values is 1.3875, which is very close to the value obtained using the central estimate.

It is possible to obtain a better estimate by taking the second point to be closer to P than this one.

Note that in this case the backward difference will give a value that is too small and the forward difference will give a value that is too large.

The central estimate will usually give a more accurate result than either of the others and will be between them.

Worked example

Improving the estimate

(i) Find an estimate for the gradient of the curve $y = 3^x$ at $x = 3$ using the points where $x = 2$ and $x = 4$.

(ii) Use the points where

(a) $x = 2.5$ and $x = 3.5$,

(b) $x = 2.9$ and $x = 3.1$.

to find improved estimates for the gradient at $x = 3$.

Solution

(i) The points on the curve are (2, 9) and (4, 81)

giving gradient $= \dfrac{81 - 9}{4 - 2} = 36.$

(ii) (a) The points are (2.5, 15.59) and (3.5, 46.76)

giving gradient $= \dfrac{46.76 - 15.59}{3.5 - 2.5} = 31.17.$

(b) The points are (2.9, 24.19) and (3.1, 30.13)

giving gradient $= \dfrac{30.13 - 24.19}{3.1 - 2.9} = 29.72.$

Exam-style question

TESTED

You are given that the equation of a curve is $y = \frac{1}{4}3^x$.

(i) Sketch a graph of the curve in the range $0 \leqslant x \leqslant 4$.

(ii) Add to your sketch the tangent to the curve at $x = 3$.

(iii) Using points on the curve when $x = 2$ and $x = 4$ find an estimate for the gradient of the tangent at $x = 3$.

(iv) Without doing any further calculations, explain how an improved accuracy can be achieved for the gradient of this tangent.

Short answer on page 128

Full worked solution online

CHECKED ANSWERS

13.4 The area under a curve

REVISED

Key facts

1 Area under a graph

The term 'area under a graph' usually means the area between the x-axis, the curve and two vertical lines.

If the curve is below the x-axis then the calculation of area gives a negative answer.

If the curve cuts the x-axis between the limits of the region, then the region should be split into two parts. Calculation of the area of one part will give a negative answer and of the other part a positive answer.

2 The rectangle rule

The area is split into rectangles of equal width. The number of rectangles or strips is usually denoted by n.

The height of each rectangle can either be the y-coordinate at the left-hand side of the rectangle or the right-hand side.

The height of the rectangle can also be the y-coordinate of the midpoint of the strip.

3 The trapezium rule

A more accurate estimate of the area of each strip is found by treating it as a trapezium.

If the first and last points on the curve are (x_0, y_0) and (x_n, y_n), then the trapezium rule is

$$A = \frac{1}{2}h\left(y_0 + 2\left(y_1 + y_2 + \ldots + y_{n-1}\right) + y_n\right)$$

where $h = \frac{x_n - x_0}{n}$.

4 Improving the accuracy

The accuracy of an estimate for area can usually (but not always) be improved by increasing the number of strips.

Note: Area is actually a positive quantity. A negative value is a warning that the region is below the x-axis.

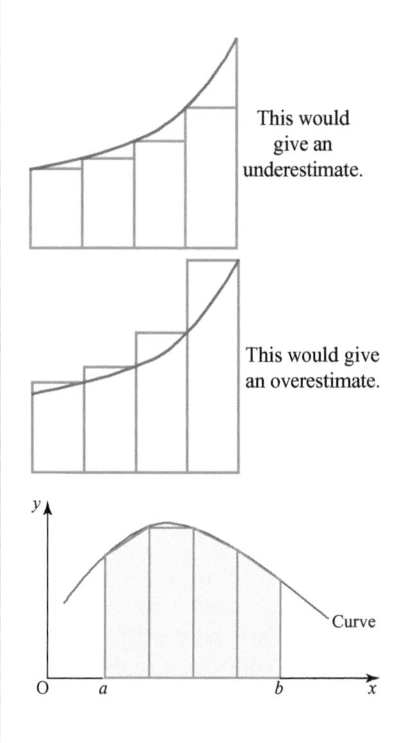

Note: Using calculus, you can calculate the area under the curves of some functions (e.g. polynomials) exactly. This is covered in Chapter 15 on integration.

Worked example

Using rectangles

1 It is required to estimate the area between the curve $y = 1 + 9x - x^2$, the x-axis and the lines $x = 1$ and $x = 3$ using 4 strips.

(i) Use four rectangles below the curve.

(ii) Use four rectangles above the curve.

(iii) Comment on the accuracy of your answers.

Solution

The interval of 2 units is split into 4 strips, so each strip is 0.5 units wide.

(i) Take y_0 as the height of the first strip

x	1	1.5	2	2.5
y	9	12.25	15	17.25
Area	4.5	6.125	7.5	8.625

Sum of rectangles = 26.75

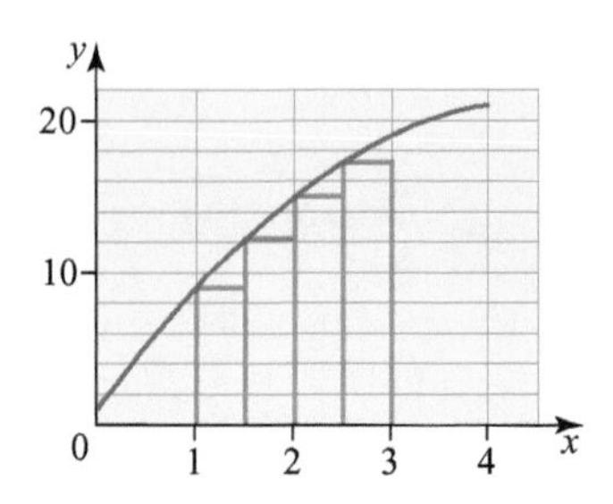

(ii) Take y_3 as the height of the last strip

x	1.5	2	2.5	3
y	12.25	15	17.25	19
Area	6.125	7.5	8.625	9.5

Sum of rectangles = 31.75

(iii) One is an overestimate and the other an underestimate. A better estimate therefore is the mean of the two, that is, 29.25.

Worked example

Using trapezia

2 Use the trapezium rule with 4 trapezia to estimate the area in example 1.

Solution

$h = 0.5.$

$$A \approx \frac{1}{2}h\left(y_0 + 2\left(y_1 + y_2 + y_3\right) + y_4\right)$$

$$= \frac{1}{4}\left(9 + 19 + 2\left(12.25 + 15 + 17.25\right)\right)$$

$$= \frac{1}{4} \times 117 = 29.25$$

Note that this answer is midway between the two answers using rectangles and is likely to be a much better approximation.

Exam-style question

TESTED

A function f(x) has values given in the following table.

x	0	1	2	3
f(x)	13	17	13	7

(i) On a grid plot the points and draw a smooth curve through them.

(ii) Using the trapezium rule with 3 trapezia find an estimate for the area between the curve $y = \mathrm{f}(x)$, the x-axis and the lines $x = 0$ and $x = 3$.

(iii) Comment on whether this is an underestimate or an overestimate or whether it is not possible to tell.

Short answer on page 128

Full worked solution online

CHECKED ANSWERS

Review questions (Chapters 12–13)

1 You are given that the points A and B lie on a curve whose equation is $y = ka^x$.
The coordinates of A and B are (2, 4.5) and (3, 13.5) respectively.
Find the values of a and k. **[3]**

2 If a sum of money is placed in a savings account which pays interest of 5% per year, paid at the end of the year, how many years does it take for the amount of money to be doubled? **[3]**

3 Solve the equation $\log_{10} x - \log_{10} 2 = 2$. **[3]**

4 Show that $3\log_2 4 - 5\log_2 8 = k$, where k is an integer to be determined. **[3]**

5 Melissa has collected some data giving values of y for integers x. She believes that the data can be modelled by the equation $y = ka^x$.

(i) Show how this proposal can be tested by plotting a straight-line graph of $\log y$ against x. **[2]**

The data she has collected is given in the table below.

x	2	3	4	5	6
y	1.7	2.5	3.8	4.7	8.5

One of the values for y does not in fact fit this model.

(ii) Plot $\log y$ against x. **[4]**

(iii) State which value of y does not fit this model. **[1]**

(iv) Draw a line of best fit through the other 4 points. Hence find values for a and k. **[4]**

6 It is thought that the data given in the table can be modelled by the equation $y = kx^n$.

x	1	2	3	4	5	6
y	0.5	4	13.5	32	62.5	108

(i) Show how a straight line of points can be drawn by taking logarithms. **[2]**

(ii) Plot $\log y$ against $\log x$ and draw the line of best fit. **[4]**

(iii) Hence find the values of k and n. **[2]**

7 Solve the equation $3^{x+1} = 2^{2x-1}$ giving your answer correct to three significant figures. **[3]**

8 Susie invests £5000 in a savings account. The account pays interest of 3% at the end of each year into the account. After how many years does the account exceed £7000?

Short answers on pages 128–29
Full worked solutions online

CHECKED ANSWERS

Section 6 Calculus

Target your revision (Chapters 14–16)

Try answering each question below. If you get stuck, follow the page reference underneath to revise that topic.

1 **Differentiation**
Differentiate $y = 2x^2 + \frac{1}{2}x$.
(see page 93)

2 **Gradients of curves**
Find the gradient of the curve $y = x^3 - 2x^2 + 3x + 1$ at the point (1, 3).
(see page 93)

3 **Tangents**
Find the equation of the tangent to the curve $y = x^3 + x^2 - 4x + 3$ at the point (2, 7).
(see page 95)

4 **Normals**
Find the equation of the normal to the curve $y = x^3 + x^2 - 4x + 3$ at the point (2, 7).
(see page 95)

5 **Stationary points and the second derivative**
Find the coordinates of the stationary points of the curve $y = x^3 + 3x^2 - 3x + 1$ and identify their nature.
(see page 96)

6 **Integration**
Find $\int (3x^2 - 3x + 1)\,dx$.
(see page 99)

7 **The equation of a curve given its gradient function and a point through which it passes.**
The gradient function of a curve is given by $\frac{dy}{dx} = 1 + 2x - 4x^3$. Find the equation of the curve given that it passes through the point (2, 5).
(see page 99)

8 **Definite integrals**
Find $\int_{-1}^{3} (x^2 + 1)\,dx$.
(see page 100)

9 **Area between a curve and the x-axis**
Find the area between the lines $x = 1$, $x = 3$, the x-axis and the curve $y = 8x - x^2$.
(see page 100)

10 **Area between two curves**
Find the area enclosed by the curves $y = x^2 + x - 1$ and $y = 9x - x^2 - 7$.
(see page 100)

11 **Constant acceleration**
On leaving a speed limit, a car accelerates uniformly from 15 ms^{-1} to 30 ms^{-1} in 250 m.
(i) Find the acceleration.
(ii) Find the time to cover the 250 m.
(see page 105)

12 **Kinematics and graphs**
A car moves from rest and accelerates for 10 seconds to reach a speed of 15 ms^{-1}. The velocity-time graph is shown here.

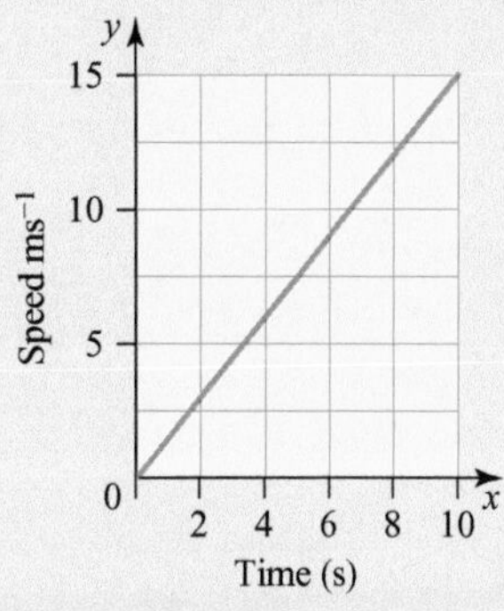

(i) How can you tell from the graph that the acceleration is constant?
(ii) Find the value of the acceleration.
(iii) Find the distance travelled by the car during the 10 seconds.
(see page 105)

13 **Acceleration due to gravity**
A ball is thrown upwards at 18 ms^{-1}. Find
(i) the greatest height reached,
(ii) the time taken before it returns to the height from where it was thrown correct to 1 decimal place.
(see page 105)

14 **Variable acceleration (1)**
A particle moves in a straight line from rest away from at point O. At time t seconds the velocity in ms^{-1} is given by $v = 4t - t^2$.
(i) Find the acceleration when $t = 5$. Explain the significance of your answer.
(ii) Find the displacement when $t = 5$.
(iii) Find the time when the particle is once again at the point O.
(see page 107)

15 Variable acceleration (2)

A train, initially at rest in a station moves with velocity v ms^{-1} at time t seconds where $v = 0.09t^2 - 0.002t^3$ until the acceleration first reaches 0 ms^{-2} after which it travels at constant speed.

(i) Find the time taken for the train to reach the point where the acceleration is zero and the speed at which it is then travelling.

(ii) Find the distance travelled by this time.

(see page 107)

Short answers on page 129

Full worked solutions online

CHECKED ANSWERS

Chapter 14 Differentiation

About this topic

Differentiation is the start of one of the really big ideas of mathematics, namely calculus.

It allows you to use the equation of a curve to find its gradient at any point.

In this chapter this is applied to finding the equations of tangents and normals, and to determining the turning points of a curve.

Before you start, remember ...

- how to find the equation of a straight line
- how to sketch curves of the form $y = f(x)$ where $f(x)$ is a quadratic or cubic polynomial
- the relationship between the gradients of tangents and normal.

14.1 Differentiation and gradients of curves

REVISED

Key facts

1 **The gradient of a curve at a particular point**

The gradient of a curve at a point is the gradient of the tangent at that point.

2 **Finding the gradient function**

The gradient function, denoted $\frac{dy}{dx}$, is used to find the gradient of a curve.

3 **Finding the gradient of a curve at a point**

The gradient of the curve at a point $x = x_1$ is found by substituting $x = x_1$ into the gradient function.

4 **Differentiating**

For a curve $y = ax^n$ where a is a constant and n is a positive integer, $\frac{dy}{dx} = nax^{n-1}$

The process is called differentiation and the result is sometimes called the derivative.

5 **The sums and differences of terms**

Each term of an expression is treated separately.
The gradient function of a constant is 0.

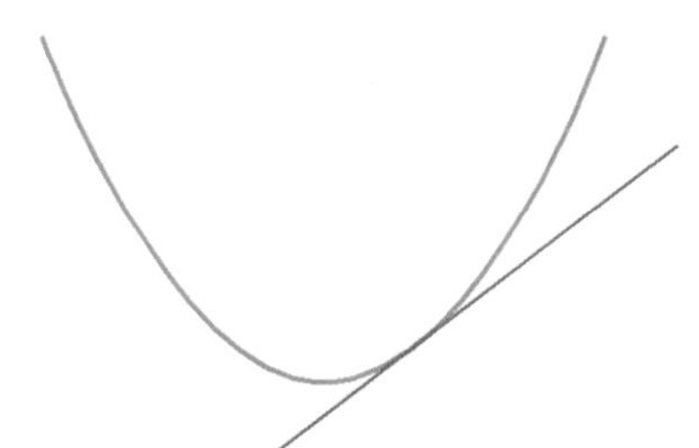

Worked example

Finding the gradient function

1 Each term is differentiated separately.

Find the gradient function of the curve $y = x^3 + 2x^2$.

Solution

$$y = x^3 + 2x^2$$

$$\Rightarrow \frac{dy}{dx} = 3x^2 + 4x$$

There are two terms here. Each should be differentiated separately.

Apply the rule for differentiation to each term. In the first term $a = 1$ but in the second $a = 2$.

Worked example

Finding the gradient of a curve at a point

2 Find the gradient of the curve $y = 3x^3 - 2x^2$ at the point (1, 1).

Solution

$$y = 3x^3 - 2x^2$$

$$\Rightarrow \frac{dy}{dx} = 9x^2 - 4x$$

When $x = 1$, $\frac{dy}{dx} = 9 - 4 = 5$

Differentiate each term separately

Substitute $x = 1$ into the gradient function to find the gradient at a particular point.

Worked examples

Finding the coordinates of a point with a given gradient

In the following example you work backwards to find x when $\frac{dy}{dx}$ is a quadratic expression.

3 Find the coordinates of the point on the curve $y = x^2 - 3x + 5$ at which the gradient is 1.

Solution

$$y = x^2 - 3x + 5$$

$$\Rightarrow \frac{dy}{dx} = 2x - 3$$

When $\frac{dy}{dx} = 1$, $2x - 3 = 1$

$\Rightarrow 2x = 4 \Rightarrow x = 2$

$y = 2^2 - 3 \times 2 + 5 = 3$

i.e. coordinates are $(2, 3)$

Differentiate each term. Remember that 5 differentiated is 0.

Set $\frac{dy}{dx} = 1$ and solve the resulting equation in x.

Don't forget to find the y-coordinate as well!

The next example involves a cubic function; there are two points with the same gradient.

4 Find the points on the curve $y=2x^3+3x^2-10x+3$ at which the gradient is 2.

Solution

$$y=2x^3+3x^2-10x+3$$

$$\Rightarrow \frac{dy}{dx}=6x^2+6x-10$$

Differentiate term by term.

$$\text{When } \frac{dy}{dx}=2,\ 6x^2+6x-10=2$$

Set $\frac{dy}{dx}=2$. Because the curve is a cubic the gradient function will be a quadratic.

$$\Rightarrow 6x^2+6x-12=0$$

$$\Rightarrow x^2+x-2=0$$

Divide by the common factor of 6.

$$\Rightarrow (x+2)(x-1)=0$$

Factorise the resulting quadratic.

$\Rightarrow$ Either $x=1$ and so $y=2+3-10+3=-2$

or $x=-2$ and so $y=-16+12+20+3=19$

For each x-coordinate find the y-coordinate.

The coordinates are $(1,-2)$ and $(-2,19)$

Exam-style question

TESTED

The equation of a curve is $y=x^3+3x^2-3x+7$.

(i) Find the gradient of the curve when $x=2$.

(ii) Show that there is no point on the curve at which $\frac{dy}{dx}=-7$.

Short answer on page 129
Full worked solution online

CHECKED ANSWERS

14.2 Tangents and normals

REVISED

Key facts

1 **Finding the gradient of a tangent**

The gradient of the tangent at a point (x_1, y_1) on a curve, is the value of $\frac{dy}{dx}$ at that point.

If that value is m then the equation of the tangent is $y-y_1=m(x-x_1)$.

2 **Finding the gradient of a normal**

The normal to the curve at a point is the line through the point perpendicular to the tangent. If the gradient of the tangent is m_1 and the gradient of the normal is m_2 then $m_1 \times m_2=-1$.

Worked example

The tangent to a curve

1 Find the equation of the tangent to the curve $y=x^2-3x+4$ at the point (1, 2).

Solution

$$y=x^2-3x+4$$

$$\Rightarrow \frac{dy}{dx}=2x-3$$

When $x=1$, $\frac{dy}{dx}=2\times1-3=-1$

Find the gradient function and then the gradient at that point.

The tangent is the line through (1, 2) with gradient $m=-1$

Use the standard form for the equation of a straight line through a given point with given tangent.

$y-2=-1(x-1)$

that is,

$\Rightarrow x+y=3$

Note that the equation of a line should contain only three terms.
The answer can also be written as $y=-x+3$.

Worked example

The normal to a curve

2 A 'normal' is a line through the point of intersection of the tangent with the curve and is perpendicular to it.

Find the equation of the tangent to the curve $y=x^2-4x+7$ at the point (2, 2).

Solution

$$y=x^2-4x+7$$

$$\Rightarrow \frac{dy}{dx}=2x-4$$

When $x=3$, $\frac{dy}{dx}=2\times3-4=2$

that is, the gradient of the tangent at (3, 4) is 2

Find the gradient of the tangent.

So the gradient of the normal at that point is $-\frac{1}{2}$

Use $m_1m_2=-1$ to find the gradient of the normal.

So the gradient of the normal is $y-4=-\frac{1}{2}(x-3)$

Then find the equation of the normal in the usual way, simplifying it to three terms.

$\Rightarrow 2y-8=3-x$

$\Rightarrow 2y+x=11$

Exam-style question

TESTED

A curve has equation $y=x^2-4x-1$ and a line has equation $2y-x+9=0$. The curve and the line meet at two points.

(i) Find the coordinates of the two points of intersection.
(ii) Determine whether the line is a normal to the curve at either point.

Short answer on page 129
Full worked solution online

CHECKED ANSWERS

14.3 Stationary points and the second derivative

REVISED

Key facts

1 **Stationary points**

A stationary point on a curve is a point where the gradient is zero. A stationary point is sometimes called a turning point.

2 **The nature of turning points**

There are three types of stationary points:

- maximum
- minimum
- stationary point of inflection.

Maxima and minima are often called turning points.

3 **Sketching curves**

When sketching a curve the essential details should be shown. This includes the intercepts on the axes and, if you know the coordinates of any turning points then these should also be included.

4 **Determining the nature of a turning point**

Two ways of deciding on the nature of a turning point are:

(i) to use values of the function

(ii) to use values of the gradient.

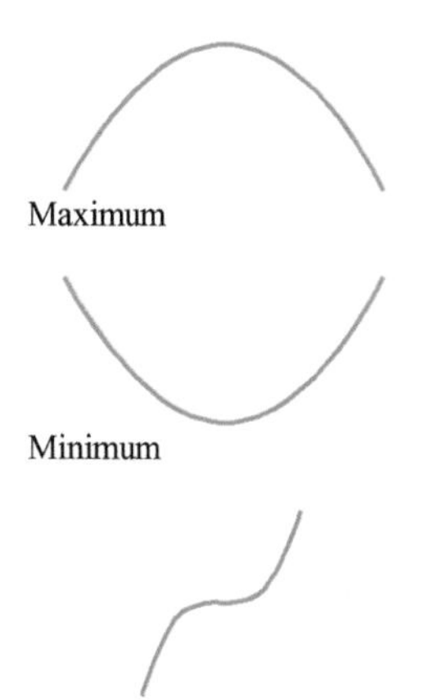

Extension: A third way to determine the nature of a turning point is to use $\frac{d^2y}{dx^2}$.

Worked example

Finding turning points

1 Find the coordinates of the turning points of the curve $y = 2x^3 - 3x^2 - 12x + 24$.

Solution

$y = 2x^3 - 3x^2 - 12x + 24$

$\Rightarrow \frac{dy}{dx} = 6x^2 - 6x - 12$

Differentiate.

$= 0$ when $6x^2 - 6x - 12 = 0$

Set $\frac{dy}{dx}$ equal to 0. Note that a common factor is extracted here, making the algebra a little easier.

$\Rightarrow x^2 - x - 2 = 0 \Rightarrow (x-2)(x+1) = 0$

So either $x = 2$, giving $y = 16 - 12 - 24 + 24 = 4$

or $x = -1$, giving $y = 31$

$\Rightarrow$ Coordinates of the turning points are (2, 4) and (–1, 31)

Worked example

Using values of the function

2 Determine the nature of the turning point (2, 4) of the curve in Example 1 above by using values of the function.

Solution

$f(1.9) = 4.088$, $f(2) = 4$, $f(2.1) = 4.092$

Try values either side of $x = 2$, so 1.9 and 2.1.

So the turning point at (2, 4) is a minimum.

Worked example

Use values of the gradient

3 Determine the nature of the other turning point, (−1, 31), of the curve in Example 1 on the previous page by using values of the gradient.

Solution

$$y = 2x^3 - 3x^2 - 12x + 24$$

$$\Rightarrow \frac{dy}{dx} = 6x^2 - 6x - 12$$

The points with $x = -1.1$ and $x = -0.9$ are either side of the turning point -1, 31).

r	-1.1	-1	-0.9
$\frac{dy}{dx}$	1.86	0	-1.74

Since the gradient goes from +ve to 0 to -ve, the turning point is a maximum.

For a maximum point, the gradient changes from having a positive value, through 0 to having a negative value.

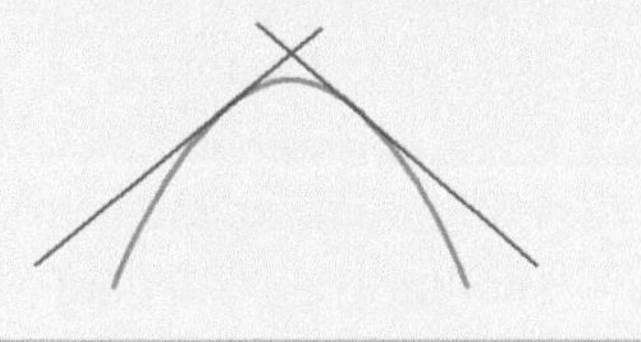

For a minimum point the gradient changes from having a negative value, through 0 to having a positive value.

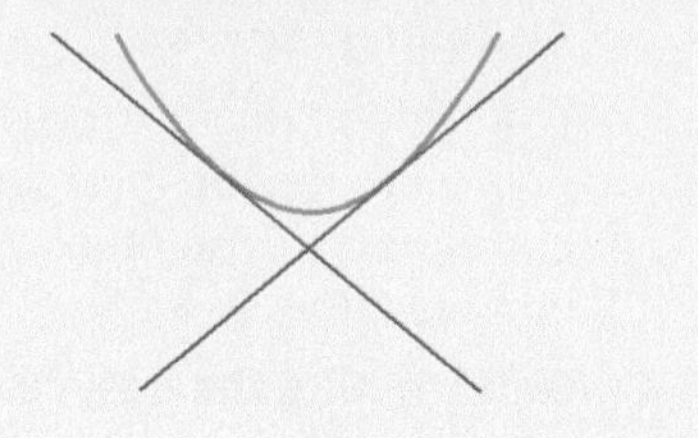

Worked example

Use the second derivative

4 Determine the nature of both turning points of Example 1 on the previous page using the second derivative.

Solution

$$y = 2x^3 - 3x^2 - 12x + 24$$

$$\Rightarrow \frac{dy}{dx} = 6x^2 - 6x - 12$$

$$\Rightarrow \frac{d^2y}{dx^2} = 12x - 6$$

When $x = 2$, $\frac{d^2y}{dx^2} = 24 - 6 = 18 > 0$

so this turning point is a minimum.

When $x = -1$, $\frac{d^2y}{dx^2} = -12 - 6 = -18 < 0$

so this turning point is a maximum.

$\frac{d^2y}{dx^2}$ measures the rate of change of $\frac{dy}{dx}$.

If $\frac{d^2y}{dx^2} < 0$ the turning point is a maximum.

If $\frac{d^2y}{dx^2} > 0$ the turning point is a minimum.

Worked example

Sketching the curve

5 Sketch the curve given in Example 1 on the previous page.

Solution

From the examples above you know that:

- there is a minimum at (2, 4)
- a maximum at (−1, 31)
- the curve also goes through (0, 24).

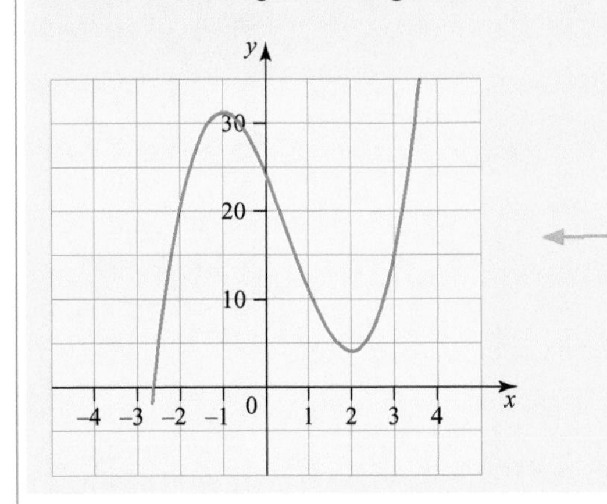

Since the minimum point is above the x-axis the curve only passes through the x-axis once, to the left of $x = -1$.

Exam-style question

TESTED

The curve $y = x^3 - 3x^2 - 9x + 7$ has two turning points, one of which is where $x = 3$.

(i) Find the coordinates of the other turning point.

(ii) Determine whether this is a maximum or minimum point.

(iii) Sketch the curve.

Short answer on pages 129–30

Full worked solutions online

CHECKED ANSWERS

Chapter 15 Integration

About this topic

Integration is the opposite process to differentiation. You can therefore find the equation of a curve if you know the gradient function. The technique is also useful for finding the area under a curve.

Before you start, remember ...

- differentiation.

15.1 Integration

REVISED

Key facts

1 **Integration is the reverse of differentiation**

If $\frac{dy}{dx} = ax^n$ then $y = \frac{ax^{n+1}}{n+1} + c$.

c is called the constant of integration.

2 **Notation**

The notation for integration is $\int f(x)dx$.

Read this as 'the integral of f(x) with respect to x'. The 'dx' should not be separated from the integral sign.

3 **Indefinite integrals**

If $\frac{d}{dx}(g(x)) = f(x)$ then $\int f(x)dx = g(x) + c$.

In an indefinite integral no limits are given so the answer includes a constant of integration.

4 **Equation of a curve**

If you are given the gradient function of a curve and the coordinates of a point on it, you can use integration to find the equation of the curve.

5 **Family of curves**

If you are given the gradient function of a curve but do not know the coordinates of any points on it, then integration will give you the equation of the curve subject to a constant of integration. So you will have a family of curves.

In this course n is 0 or a positive integer.

If you differentiate a constant, c, you get 0, so when you integrate a function you have to write it in.

Worked examples

Indefinite integrals

1 Integrate $3x^4+2x^5$ with respect to x.

Solution

The '$\mathrm{d}x$' means integrating with respect to x. It is a crucial part of the notation and should never be omitted.

$$\int(3x^4+2x^5)\,\mathrm{d}x=\frac{3x^5}{5}+\frac{x^6}{3}+c$$

Integrate, using the rule for integration, term by term. Don't forget the constant of integration, c.

2 Given that $\frac{\mathrm{d}y}{\mathrm{d}x}=4x^2+x^3-5x^4$, find y.

Solution

$$\frac{dy}{dx}=4x^2+x^3-5x^4$$

$$\Rightarrow y=4\times\frac{x^3}{3}+\frac{x^4}{4}-5\times\frac{x^5}{5}+c$$

$$\Rightarrow y=\frac{4x^3}{3}+\frac{x^4}{4}-x^5+c$$

Integrate term by term and add the arbitrary constant of integration, c.

Worked example

The equation of a curve

3 The gradient function of a curve is given by $\frac{\mathrm{d}y}{\mathrm{d}x}=3x^2-2x+1$. Find the equation of the curve given that it passes through the point (1, 2).

Solution

$$\frac{dy}{dx}=3x^2-2x+1$$

$$\Rightarrow y=x^3-x^2+x+c$$

Integrate to get the equation of the family of curves.

Through $(1, 2)\Rightarrow 2=1-1+1+c$

Now substitute (1, 2) to find the particular equation.

$$\Rightarrow c=1$$

$$\Rightarrow y=x^3-x^2+x+1$$

Exam-style question

TESTED

(i) Find $\int(2x+2)\,\mathrm{d}x$.

(ii) A family of curves have gradient given by $\frac{\mathrm{d}y}{\mathrm{d}x}=2x+2$. Find the equation of the family of curves.

(iii) Sketch three curves in the family of curves.

(iv) Find the equation of the curve that passes through the point (3, 0).

Short answer on page 130

Full worked solution online

CHECKED ANSWERS

15.2 Definite integrals and area

REVISED

Key facts

1 Definite integrals

A definite integral has a numerical value and is obtained from the indefinite integral by substituting the two limits and subtracting.

If $\int f(x)dx = g(x)+c$ then $\int_a^b f(x)dx = [g(x)]_a^b = g(b)-g(a)$.

Note that the arbitrary constant of integration, c, is not involved.

2 Area under a curve

The area between the curve $y=f(x)$, the lines $x=a$, $x=b$ and the x-axis is $\int_a^b f(x)dx$.

3 Properties of the area

Integration gives a sign to the area under a curve.

- If the curve $y=f(x)$ lies entirely above the x-axis then integration gives a positive answer.
- If the curve $y=f(x)$ entirely below the x-axis then integration gives a negative answer.
- If the curve $y=f(x)$ crosses the x-axis between $x=a$ and $x=b$ then the integration needs to be carried out in two parts; one will give a positive answer and the other a negative answer.

Note: Area is actually a positive quantity. A negative value is a warning that the region is below the x-axis.

4 Area between two curves

If the curves $y=f(x)$ and $y=g(x)$ meet at $x=a$ and $x=b$ and within this region $f(x) > g(x)$ then the area enclosed between the curves is $\int_a^b f(x)dx - \int_a^b g(x)dx = \int_a^b (f(x)-g(x)) dx$.

Note that it is possible to find the first definite integral and then the second and subtract, or to subtract before integrating.

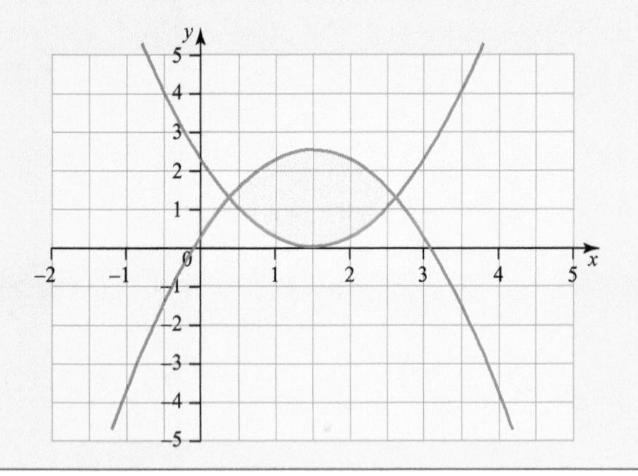

Worked examples

Definite integrals

1 (i) Find the area between the curve $y = 3 + 4x - x^2$, the x-axis and the lines $x = 1$ and $x = 4$.
(ii) Sketch the graph of the curve and lines, shading the region whose area you have found.

Solution

(i)

$$\text{Area} = \int_1^4 (3 + 4x - x^2)\,dx = \left[3x + 2x^2 - \frac{x^3}{3}\right]_1^4$$

Integrate term by term. Ignore the arbitrary constant of integration

$$= \left(3 \times 4 + 2 \times 4^2 - \frac{4^3}{3}\right) - \left(3 + 2 - \frac{1}{3}\right)$$

$$= \left(44 - \frac{64}{3}\right) - \left(5 - \frac{1}{3}\right) = 39 - \frac{63}{3}$$

Substitute the upper limit and the lower limit and subtract.

$$= 39 - 21 = 18$$

(ii)

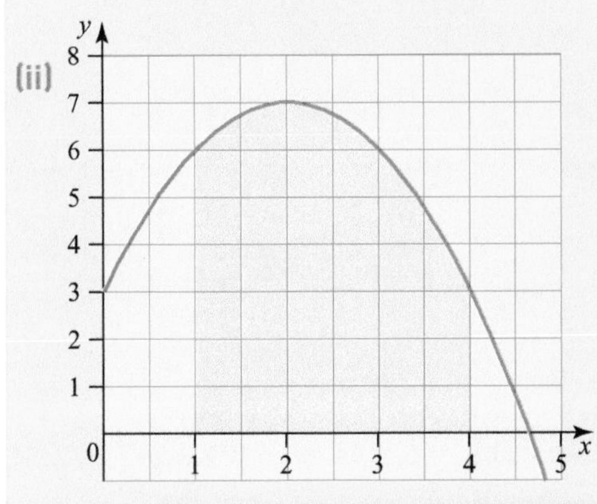

2 (i) Find the positive area between the curve $y = 2x - x^2$ and the x-axis.
(ii) Sketch the curve and shade the region whose area you have found.

Solution

(i) The curve cuts the x-axis at $x = 0$ and $x = 2$.

$$\text{So Area} = \int_0^2 (2x - x^2)\,dx = \left[x^2 - \frac{x^3}{3}\right]_0^2$$

$$= \left(2^2 - \frac{2^3}{3}\right) - (0) = 4 - \frac{8}{3} = \frac{4}{3}.$$

Find the points where the curve cuts the x-axis by solving the equation $2x - x^2 = 0$.
Integrate and substitute limits.

(ii)

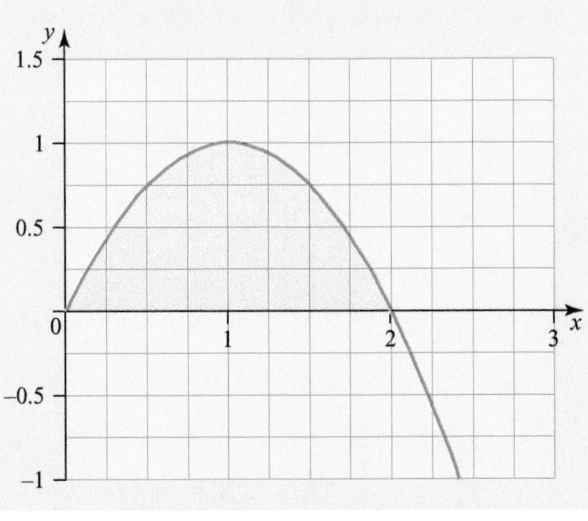

Worked example

Area between two curves

3 The graph shows the curves $y = 13 - 6x + x^2$ and $y = 3 + 6x - x^2$.

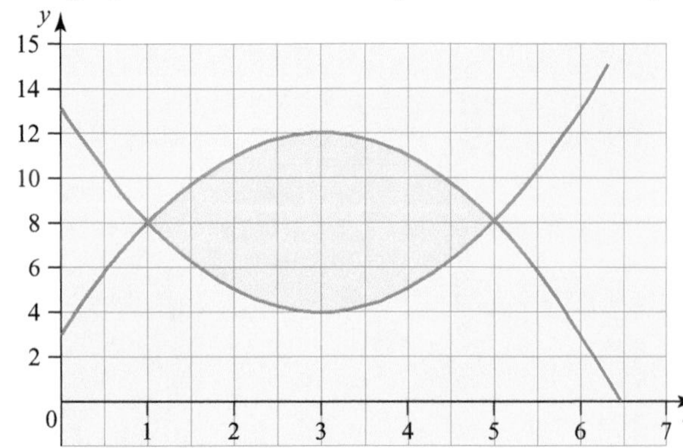

Find the area enclosed between them.

Solution

We first need to find where the curves intersect.

$$13 - 6x + x^2 = 3 + 6x - x^2$$

$$\Rightarrow 2x^2 - 12x + 10 = 0$$

$$\Rightarrow \quad x^2 - 6x + 5 = 0 \Rightarrow (x-1)(x-5) = 0$$

You only need the x coordinates.

$$\Rightarrow \quad x = 1 \text{ and } x = 5$$

This method evaluates each integral separately and subtracts. This is effectively finding the area under each curve with the area between them being the difference.

Integrate and substitute the limits.

Method 1

Area under 'top' curve

$$= \int_1^5 (3 + 6x - x^2)\,dx = \left[3x + 3x^2 - \frac{x^3}{3}\right]_1^5$$

$$= \left(15 + 75 - \frac{125}{3}\right) - \left(3 + 3 - \frac{1}{3}\right)$$

$$= 48\frac{1}{3} - 5\frac{2}{3} = 42\frac{2}{3}$$

Area under 'bottom' curve

Repeat for the lower curve.

$$= \int_1^5 (13 - 6x + x^2)\,dx = \left[13x - 3x^2 + \frac{x^3}{3}\right]_1^5$$

$$= \left(65 - 75 + \frac{125}{3}\right) - \left(13 - 3 + \frac{1}{3}\right)$$

$$= 31\frac{2}{3} - 10\frac{1}{3} = 21\frac{1}{3}$$

$$\text{Total area} = 42\frac{2}{3} - 21\frac{1}{3} = 21\frac{1}{3}$$

Subtract the two values to give final answer.

This method subtracts the functions and then integrates. Note that if you do the subtraction the wrong way around your answer will come out to be negative.

Method 2

$$\text{Area} = \int_1^5 \left(\left(3+6x-x^2\right)-\left(13-6x-x^2\right)\right)dx$$

Combine the two curves by subtracting.

$$= \int_1^5 \left(3-13+6x+6x-x^2-x^2\right)dx$$

$$= \int_1^5 \left(-10+12x-2x^2\right)dx = \left[-10x+6x^2-\frac{2x^3}{3}\right]_1^5$$

$$= \left(-50+150-\frac{250}{3}\right)-\left(-10+6-\frac{2}{3}\right)$$

$$= \frac{50}{3}-\left(-4\frac{2}{3}\right) = 16\frac{2}{3}+4\frac{2}{3} = 21\frac{1}{3}$$

Note that you might get these the 'wrong way round' if you do not know which is the 'upper' curve. That means that the answer will be negative, in which case drop the negative sign at the end.

Exam-style question

TESTED

The figure shows the curves with equations $y=4x-x^2$ and $y=x^2-4x+6$.

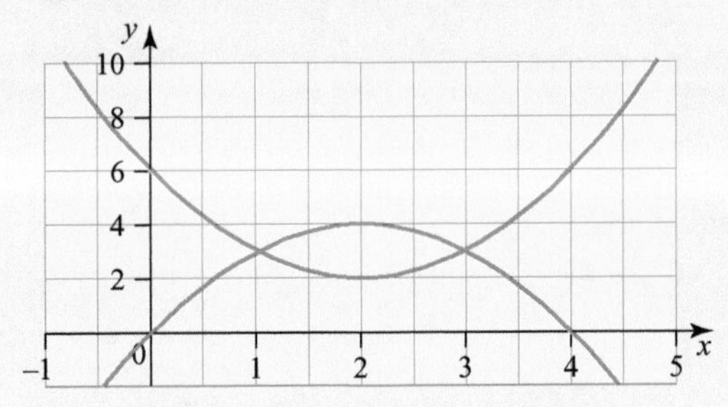

(i) Use an algebraic method to find the x-coordinates of the points where the curves intersect.
(ii) Calculate the area enclosed by the curves.

Short answer on page 130
Full worked solution online

CHECKED ANSWERS

Chapter 16 Applications to kinematics

About this topic

Kinematics is the study of motion.

The rate of change of the displacement of a body is its velocity.

The rate of change of its velocity is its acceleration.

Differentiation involves rate of change, so calculus is often used in kinematics.

However, calculus is not needed in situations where the acceleration is constant.

In this course objects are modelled as particles with mass but no size.

Before you start, remember ...

- how to differentiate terms of the form ax^n where n is a positive integer or 0
- how to integrate terms of the form ax^n where n is a positive integer or 0.

16.1 Constant acceleration

REVISED

Key facts

1 **Constant acceleration**

If acceleration is constant then you can use the 'suvat' equations below. If acceleration is not constant, then you must use calculus as in the next section.

2 **'*suvat*' equations**

$$v = u + at$$

$$s = ut + \frac{1}{2}at$$

$$s = \left(\frac{u+v}{2}\right)t$$

$$s = vt - \frac{1}{2}at^2$$

$$v^2 = u^2 + 2as$$

3 **Acceleration due to gravity**

Bodies falling freely under gravity have a constant downward acceleration denoted by $g\,\mathrm{m\,s^{-1}}$. The value of g is usually taken to be 9.8, but 10 is also sometimes used.

In an examination question, if the acceleration is constant then these formulae may be quoted and used without derivation

Warning: If the acceleration is not constant then these formulae are invalid.

Common mistake: Using *suvat* formulae when the acceleration is not constant.

In an examination you should take $g = 9.8$ unless the question says otherwise.

Worked examples

Constant acceleration

1 A car starts from rest and travels with a constant acceleration of $2\,\text{m}\,\text{s}^{-2}$ along a road.
Find
(i) the distance travelled in the first 5 seconds of motion,
(ii) the velocity at that time.

Solution

(i) In this example $a = 2$, $u = 0$.

$s = ut + \frac{1}{2}at^2$ with $t = 5$ gives $s = 0 + \frac{1}{2} \times 2 \times 5^2 = 25$.

So the displacement after 5 seconds is 25 metres.

(ii) $v = u + at$ gives $v = 0 + 2 \times 5 = 0$

So velocity is 10 metres per second.

You are given u and a. The formulae $s = ut + \frac{1}{2}at^2$ and $v = u + at$ are therefore appropriate.

2 A body moves on a straight line. When it passes a point O it is moving at $3\,\text{m}\,\text{s}^{-1}$ with constant acceleration $2\,\text{m}\,\text{s}^{-2}$.
What is its speed after it has travelled 5 metres?

Solution

Using $v^2 = u^2 + 2as$ gives

$v^2 = 9 + 2 \times 2 \times 10 = 49$

$\Rightarrow v = 7\,\text{m}\,\text{s}^{-1}$

So speed is 7 metres per second.

This is the appropriate formula here. Note that it does not contain t.

Worked example

Acceleration due to gravity

3 A ball is dropped from a window 30 metres above the ground. How long does it take for the ball to reach the ground and with what velocity does it land? (Take $g = 9.8$)

Solution

You are given that $a = 9.8$ and $u = 0$.

$s = ut + \frac{1}{2}at^2$ with $s = 30$ gives $30 = 0 + \frac{1}{2} \times 9.8 \times t^2$

$\Rightarrow t^2 = 6.12$

$\Rightarrow t = 2.47$

$v^2 = u^2 + 2as$ with $s = 30$ gives $v^2 = 2 \times 9.8 \times 30 = 588$

$\Rightarrow v = 24.25$

So velocity is 24.25 metres per second and the time is 2.47 seconds.

The constant acceleration is $9.8\,\text{m}\,\text{s}^{-2}$.

Note: Acceleration has direction. The acceleration due to gravity is downwards. If you take the positive direction as upwards (often a sensible thing to do if the question is about projecting a particle upwards) then acceleration due to gravity will be a negative quantity.

Exam-style question

TESTED

A car starts from rest and travels with constant acceleration. After 8 seconds its speed is $20\,\text{m}\,\text{s}^{-1}$.

(i) Find the acceleration of the car.
(ii) Find the distance the car has travelled after 8 seconds.
(iii) Find how long it took the car to travel the first 45 m.

Short answer on page 130
Full worked solution online

CHECKED ANSWERS

16.2 Variable acceleration

REVISED

Key facts

1 Definitions and units

Quantity	Definition	SI Unit	Symbol used
Time	Measured from a fixed origin	Second (s)	t
Distance	Distance travelled in a given time	Metre (m)	s (or x or y)
Displacement	Distance from a fixed origin	Metre (m)	s (or x or y)
Speed	Rate of change of distance	Metre per second (ms^{-1})	$v = \frac{ds}{dt}$
Velocity	Rate of change of displacement	Metre per second (ms^{-1})	$v = \frac{ds}{dt}$
Acceleration	Rate of change of velocity	Metre per second per second (ms^{-2})	$a = \frac{dv}{dt} = \frac{d^2s}{dt^2}$

In this course the direction will always be along a straight line but may be positive or negative.

2 Rates of change

The relationships between displacement, velocity and acceleration are illustrated in this diagram.

Differentiate →

Displacement Velocity Acceleration

← Integrate

Worked example

Rate of change

1 A particle is initially at the origin and moves along a straight line. At time t seconds its displacement in metres is given by $s = t^3 + 2t^2 - 6t$.

Find the acceleration of the particle when $t = 3$.

Solution

$s = t^3 + 2t^2 - 6t$

$\Rightarrow v = \frac{ds}{dt} = 3t^2 + 4t - 6$ — Differentiate to find the velocity.

$\Rightarrow a = \frac{dv}{dt} = \frac{d^2s}{dt^2} = 6t + 4$ — Differentiate again to find the acceleration.

When $t = 3$, $a = 18 + 4 = 22$ — Substitute $t = 3$.

After 3 seconds the acceleration is 22 ms^{-2}

Worked example

Using integration

2 A body moves along a straight line and is initially at the origin. Its acceleration, $a\,\mathrm{m\,s^{-2}}$, at time t s is given by

$a = 5 - t$.

Given that its initial velocity is $1\,\mathrm{m\,s^{-1}}$, find expressions for the velocity and displacement t seconds after passing O.

> **Note:** 'Initial velocity' means the velocity when $t = 0$.

Solution

$a = 5 - t$

$\Rightarrow v = \int a\,dt = 5t - \frac{t^2}{2} + c.$ ← Integrate to find a function for v.

When $t = 0, v = 1 \Rightarrow c = 1$ ← Given the initial velocity, find the value of c.

$\Rightarrow v = 5t - \frac{t^2}{2} + 1.$

$s = \int v\,dt = \frac{5t^2}{2} - \frac{t^3}{6} + t + c.$ ← Integrate again to find a function for s.

When $t = 0, s = 0 \Rightarrow c = 0$ ← $s = 0$ when $t = 0$ will find c.

$\Rightarrow s = \frac{5t^2}{2} - \frac{t^3}{6} + t$

Exam-style question

TESTED

A body moves along a straight line. As it passes a point O it is travelling at $4\,\mathrm{m\,s^{-1}}$. The acceleration t seconds after passing O is given by $a = 2 + t$.

Find

(i) the velocity after 5 seconds,

(ii) the displacement at that time.

Short answer on page 130

Full worked solution online

CHECKED ANSWERS

Review questions (Chapters 14–16)

1 A curve has equation $y = x^2 - 2x + 7$.
Find the coordinates of the point on the curve where the gradient is 4. **[3]**

2 The gradient of a curve at a point (x, y) is given by $\frac{\mathrm{d}y}{\mathrm{d}x} = 2x^3 - x^2 + 5$.
Find the equation of the normal to the curve at the point $\mathrm{P}(1, 2)$. **[5]**

3 A curve has equation $y = x^3 + 2x^2 - 5x + 5$.

(i) Find the equation of the tangent at $\mathrm{P}(1, 3)$. **[4]**

(ii) This tangent cuts the curve again at point R. Find the coordinates of R. **[4]**

(iii) Determine whether the tangent at P is a normal to the curve at R. **[3]**

4 (i) Show that there is a stationary point at $(1, 9)$ on the curve $y = x^3 - 6x^2 + 9x + 5$ and determine the nature of this stationary point. **[5]**

(ii) Find the coordinates of the other stationary point. **[2]**

(iii) Hence sketch the curve. **[2]**

5 (i) Find the stationary points on the curve $y = x^3 + \frac{3}{2}x^2 - 6x + 4$. **[4]**

(ii) Determine the nature of each point. **[4]**

(iii) Sketch the curve. **[2]**

6 The gradient function of a curve is given by $\frac{\mathrm{d}y}{\mathrm{d}x} = 2 + 2x - x^2$. Find the equation of the curve given that it passes through the point $(3, 10)$. **[4]**

7 The curve shown is part of the graph of $y = 4 - x^2$.
Calculate the area of the shaded region between this curve and the x-axis, giving your answer as an exact fraction. **[4]**

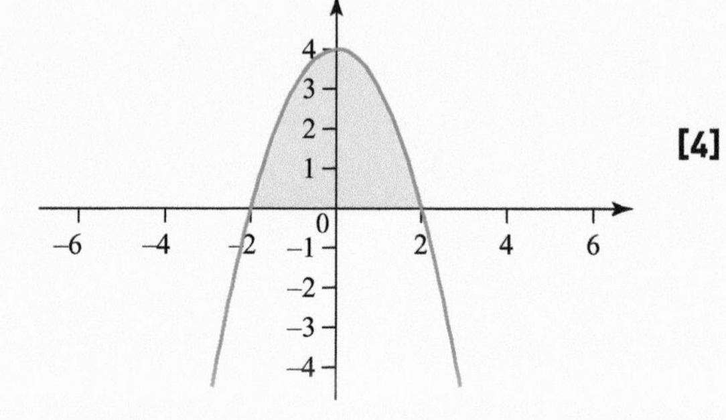

8 The graph shows the curves $y = x^2 - 7x + 8$ and $y = -x^2 + 9x - 6$.

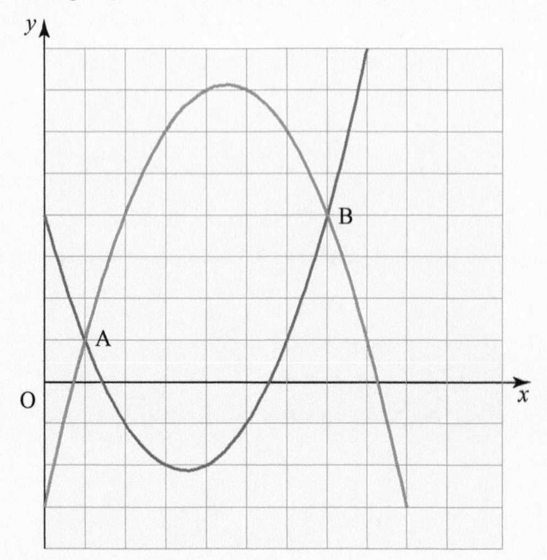

They intersect at A and B.

(i) Use an algebraic method to find the coordinates of A and B. **[4]**

(ii) Calculate the area enclosed by the two curves. **[4]**

9 A car, which is initially travelling at $20\,\mathrm{m\,s^{-1}}$, accelerates uniformly at $1.2\,\mathrm{m\,s^{-2}}$.

Find

(i) the speed after 5 seconds, **[2]**

(ii) the distance travelled in this time. **[2]**

10 Two cars, A and B, are travelling at a constant speed of $13\,\mathrm{m\,s^{-1}}$ along a straight road. The front of car B is 8 m behind the front of car A. The driver of car B decides to overtake A and so accelerates uniformly at $0.5\,\mathrm{m\,s^{-2}}$. After how many seconds is the front of car A 8 m behind the front of car B? **[5]**

11 A speedboat accelerates from rest so that t seconds after starting its velocity, in $\mathrm{m\,s^{-1}}$, is given by the formula $v = 0.36t^2 - 0.024t^3$.

(i) Find the acceleration at time t. **[3]**

(ii) Find the distance travelled in the first 10 seconds. **[4]**

Short answers on pages 130–31

Full worked solutions online

CHECKED ANSWERS

Exam preparation

Before your exam:

- Do not leave your revision until the last few weeks – little and often is the best way.
- Be organised and disciplined – don't get distracted by other things.
- Turn your phone off while revising – you won't do it if you keep talking to friends.
- Use this book to ensure you have everything covered and can understand your notes.
- Do as many questions as you can, particularly on the topics with which you do not feel comfortable.
- Do aim, however, to put aside some past and practice papers for a final check on your ability to deal with questions in the given time.
- The paper is 2 hours long and carries 100 marks. A very rough guide is a mark a minute.

Formulae you need to know

All GCSE is assumed knowledge

Polynomials and binomial expansions	**The factor theorem** If $f(a) = 0$ then $(x - a)$ is a factor of $f(x)$ and $x = a$ is a root of the equation $f(x) = 0$. **The binomial expansion** $(a+x)^n = (a)^n + {}_nC_1(a)^{n-1}(x) + {}_nC_2(a)^{n-2}(x)^2 + \ldots + {}_nC_r(a)^{n-r}(x)^r + \ldots + (x)^n$ **Factorial notation** $n! = n(n-1)(n-2)\ldots3.2.1$ **Coefficients of the binomial expansion** ${}^nC_r = {}_nC_r = \binom{n}{r} = \frac{n!}{r!(n-r)!} = \frac{n(n-1)(n-2)\ldots(n-r+1)}{1.2.3.4\ldots r}$ **Pascal's triangle** Can be used for small values for n to determine the coefficients. 1 1 1 1 2 1 1 3 3 1 1 4 6 4 1 1 5 10 10 5 1 etc.
Coordinate geometry	The equation of a straight line with gradient m and intercept, c, on the y-axis is $y = mx + c$. The equation of a straight line with gradient m and passing through the point (x_1, y_1) is $y - y_1 = m(x - x_1)$. For a line joining the points $(x_1, y_1), (x_2, y_2)$: • Distance between the points, $d = \sqrt{(x_1 - x_2)^2 + (y_1 - y_2)^2}$ • Midpoint of the line segment is $\left(\frac{x_1 + x_2}{2}, \frac{y_1 + y_2}{2}\right)$ • Gradient of the line $= \frac{y_1 - y_2}{x_1 - x_2}$ • The equation of the line is $\frac{y - y_1}{y_2 - y_1} = \frac{x - x_1}{x_2 - x_1}$
Circles	The equation of a circle, centre (a, b) and radius r is $(x - a)^2 + (y - b)^2 = r^2$

Trigonometry

Pythagoras' theorem

In a right-angled triangle where $A = 90°$

$$a^2 = b^2 + c^2$$

$$\sin^2\theta + \cos^2\theta = 1$$

$$\frac{\sin\theta}{\cos\theta} = \tan\theta$$

Sine rule

$$\frac{a}{\sin A} = \frac{b}{\sin B} = \frac{c}{\sin C} \quad \text{or} \quad \frac{\sin A}{a} = \frac{\sin B}{b} = \frac{\sin C}{c}$$

Cosine rule

$$a^2 = b^2 + c^2 - 2bc\cos A \quad \text{or} \quad \cos A = \frac{b^2 + c^2 - a^2}{2bc}$$

Calculus

$$y = ax^n \Rightarrow \frac{\mathrm{d}y}{\mathrm{d}x} = nax^{n-1} \quad \text{for } n \neq 0$$

$$y = \mathrm{f}(x) \pm \mathrm{g}(x) \Rightarrow \frac{\mathrm{d}y}{\mathrm{d}x} = \mathrm{f}'(x) \pm \mathrm{g}'(x)$$

$$\frac{\mathrm{d}y}{\mathrm{d}x} = ax^n \Rightarrow y = \frac{ax^{n+1}}{n+1} + c \quad \text{for } n \neq -1$$

Kinematics

For constant acceleration

$$s = ut + \frac{1}{2}at^2$$

$$v = u + at$$

$$v^2 = u^2 + 2as$$

For variable acceleration

$$v = \frac{\mathrm{d}s}{\mathrm{d}t}, \quad a = \frac{\mathrm{d}v}{\mathrm{d}t} = \frac{\mathrm{d}^2 s}{\mathrm{d}t^2}$$

Laws of indices

$$a^n . a^m = a^{n+m}, \quad \left(a^n\right)^m = a^{nm}, \quad \frac{a^n}{a^m} = a^{n-m},$$

$$a^{-n} = \frac{1}{a^n}, \quad a^0 = 1$$

Logarithms

$$\log x + \log y = \log xy, \quad \log x - \log y = \log\left(\frac{x}{y}\right)$$

$$\log x^n = n\log x, \quad \log 1 = 0$$

$$\log\left(\frac{1}{x}\right) = -\log x$$

During your exam

Watch out for these words

Exact... Do not use your calculator to calculate an answer. Leave it with included surds or powers.

For example, $2\sqrt{3}$ or 3π.

Show that... The answer has been given to you so you must show every step of the working.

Prove that... This is the same as 'show that...' except that it is likely to be algebraic rather than arithmetical.

Hence... You should follow a statement or a previous result.

Calculate... Unless there is a specific instruction (such as 'show all your working') you may use your calculator to do the arithmetic.

Determine... Expect to give some working to justify your result.

Give, state, write down... You should be able to write down the answer without working anything out.

Sketch, plot, draw...

- A **sketch** is a diagram that is not necessarily to scale showing the main features of the curve being drawn, such as turning points, intersections with the axes, etc. It often will involve the general shape for large positive or negative values of x.
- **Draw** means a diagram that is slightly more accurate for the context.
- **Plot** means that you have to work out values of y for a number of values of x (which will be given to you in the question) and then to mark them accurately on graph paper, joining them with a smooth curve.

Remember

If you give an answer that is correct with no working, it is possible that you will receive full marks (unless you have used a method that has been specifically prohibited). However, if you give an incorrect answer with no working then you will receive no marks at all. If you show some working, you may well receive some intermediate marks even though your final answer is wrong.

If you cannot do part (a) of a question it does not mean that you cannot do part (b), so don't ignore the rest of the question just because you cannot do the first part. If the first part is a 'show that...', then you may use this in part (b) even if you were not able to show it.

Pace yourself: 120 minutes for 100 marks means 100 minutes at a mark a minute plus 20 minutes reading and checking. If you spend 10 minutes on the first question which only carries 2 marks, then you will be putting pressure on yourself later in the exam.

Two different conventions for labelling the parts of questions have been used by OCR for this course. One way is for (a), (b) ... to be used for unrelated parts and (i), (ii) ... for parts that follow the stages through a question. This was used in this course up to 2019 and so is the style of most of the questions in this book. However, a new convention is being introduced where question parts are labelled (a), (b) ... whether they are connected or not, and the labels (i), (ii) ... are only used for their sub-parts.

If you get stuck on a question

- Don't panic!
- Move on to the next question and give yourself time to return to it.
- Re-read the question – have you missed an essential piece of information?
- Would a sketch help?

Finishing off

Try to pace yourself so that you have time to check your answers.

Check that your answer to each question is

- to the correct accuracy
- in the right form
- complete.

ANSWERS

Section 1 Algebra

Target your revision (Chapters 1–4) (pages 1–2)

1 $3x-3y+3$

2 $7\sqrt{2}$

3 $3\left(2+\sqrt{3}\right)$

4 (i) $f(x)+g(x)=x^3-x^2+4x$

(ii) $f(x)g(x)=3x^4-2x^3+2x^2-2x-1$

5 $x=4$

6 $x=-1, \frac{3}{2}$

7 $x=-1.851, 1.351$

8 $x^2-x+7=0$

Discriminant 'b^2-4ac' $=-27<0$

9 $(n-3)^2+6\geqslant 6$ for all n

10

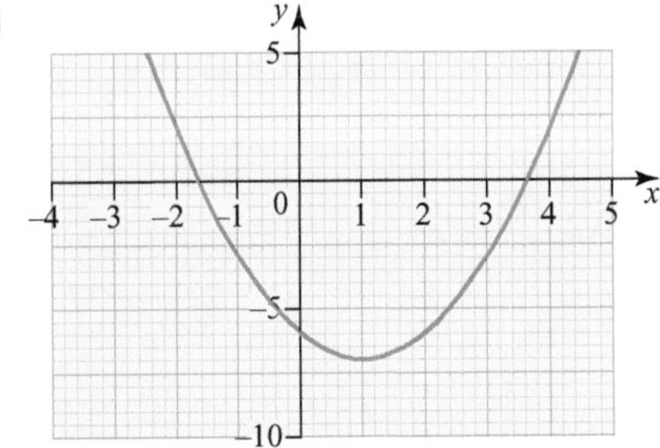

$x=3.6, -1.6$

11 $f(x)=3x^2+4x-4 \Rightarrow f(-2)=12-8-4=0$

So $(x+2)$ is a factor of f(x).

12 $x=-2, 1, 3$

13 $x=2$ and $y=3$

14 $x=2$ giving $y=3$

and $x=6$ giving $y=23$

15 John's age this year is 26 and Paul's age is 51.

16 $2<x<10$

17

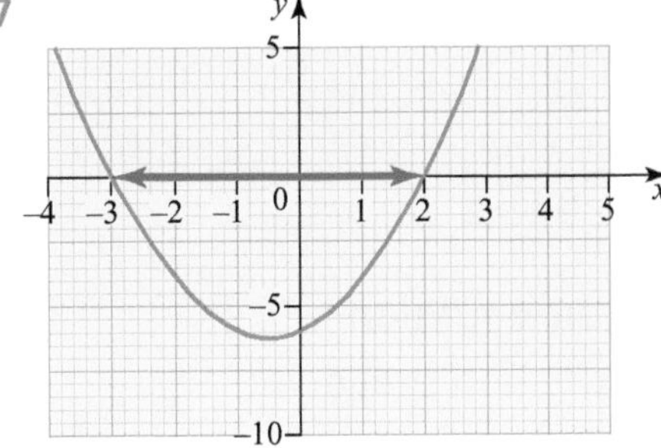

$-3\leqslant x\leqslant 2$

18 $u_1=1, u_2=3, u_3=7, u_4=15, u_5=31$

19 $u_1=2, u_2=5, u_3=10, u_4=17, u_5=26$

20 $u_1=1, u_2=2, u_3=5, u_4=12, u_5=29$

21 16 300

Chapter 1 Algebraic manipulation

1.1 Manipulating algebraic expressions Exam-style question (page 5)

$\frac{x^2-3x+10}{5x}$

1.2 Manipulating expressions involving square roots Exam-style question (page 7)

$\frac{9}{7}-\frac{4}{7}\sqrt{2}$

Chapter 2 Polynomials

2.1 Polynomials Exam-style question (page 9)

(i) x^3+2x^2+2x-7

(ii) 4

2.2 Review of solving linear equations
Exam-style question (page 11)

$x = -35$

2.3 Solving quadratic equations
Exam-style question (page 13)

(i) $\left(x - \frac{3}{2}\right)^2 - \frac{13}{4}$

(ii) $x = 3.303$ or -0.303

2.4 Solving cubic equations
Exam-style question (page 14)

(i) $f(-1) = 0$

(ii) $x = -1,\ 2$ or 4

Chapter 3 Applications of equations and inequalities in one variable

3.1 Review of simultaneous equations
Exam-style question (page 16)

$x = 3, y = 9$ and $x = -1, y = 5$

3.2 Setting up equations
Exam style question (page 19)

(i) Gavin's time $= \frac{140}{v}$

Simon's time $= \frac{140}{v+5}$

(ii) Go online for full worked solution.

(iii) Gavin's time is 2 hours 46 minutes

Simon's time is 2 hours 31 minutes

3.3 Inequalities
Exam-style question (page 21)

(i) x satisfies $x \geqslant 3$ or $x \leqslant -5$

(ii)

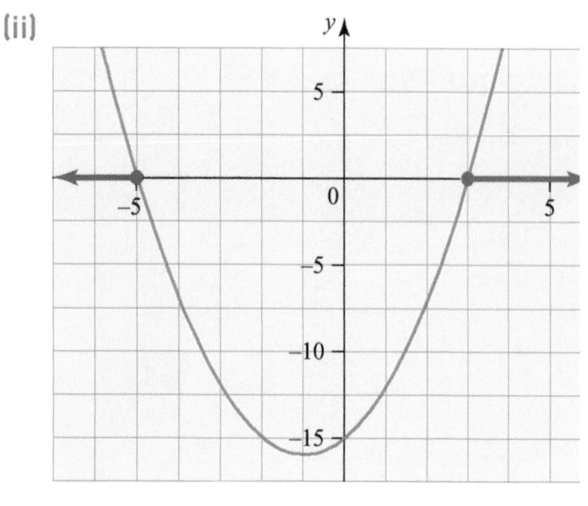

Chapter 4 Recurrence relationships

4.1 First order recurrence relationships
Exam-style question (page 24)

(i) $L_{n+1} = L_n \times 1.1$

(i) £200

(ii) £3221.02

4.2 Second order recurrence relationships
Exam-style question (page 25)

(i) $u_1 = 1$

$u_2 = 2$

$u_3 = 5$

$u_4 = 14$

$u_5 = 41$

(ii) Go online for full worked solution.

Review questions (Chapters 1–4) (page 26)

1 $2x(x+2)$

2 $-9+5\sqrt{15}$

3 2

4 $2-\frac{1}{2}\sqrt{6}$

So $a = 2, b = \frac{1}{2}, n = 6$

5 $x^2 - 3x + 9$

6 $x = -35$

7 $x = 5.32$ or -1.32

8 $x = 4 \pm \sqrt{6}$

9 (i) Go online for full worked solution.

(ii) $x = -1, 2,$ or 7

10 $a = -14,\ b = 35$

11 $x = 1, y = 4$

$x = 2, y = 5$

12 Carla's age now is 10 years.

13 (i) By day: $\frac{200}{v}$, by night $\frac{200}{v+20}$

(ii) Go online for full worked solution.

(iii) Day time, 3 hours 20 minutes

Night time, 2 hours 30 minutes

14 (i) $x < -\frac{1}{3}$

(ii) $-6 < x < 1$

15 −2, −1, 0, 1, 2, 3, 4

16 (i) 5%

(ii) £2315.25

17 (i) $B=1,\ A=-1$

(ii) $u_3=19$

$u_4=65$

Section 2 Coordinate geometry in two dimensions

Target your revision (Chapters 5–7) (page 27)

1 $6y+5x-8=0$

2 $-\dfrac{2}{5}$

3 Parallel: $3x+4y=11$
Perpendicular: $4x-3y+2=0$

4 Centre has coordinates (2, −3) and the radius is $\sqrt{16}=4$

5

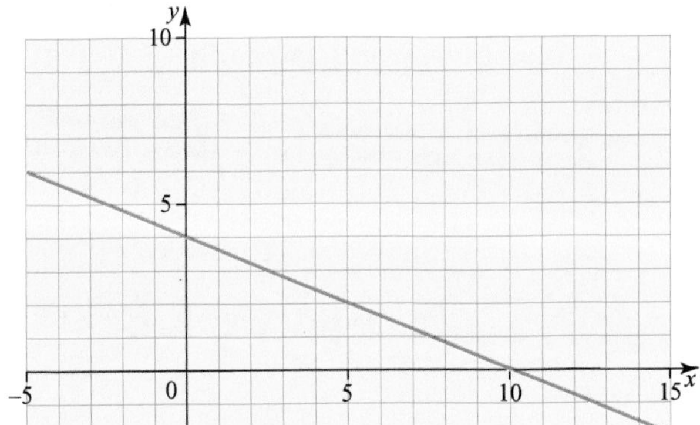

6

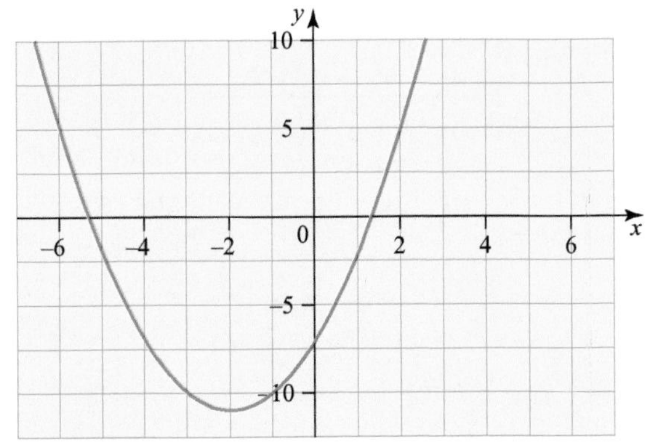

7

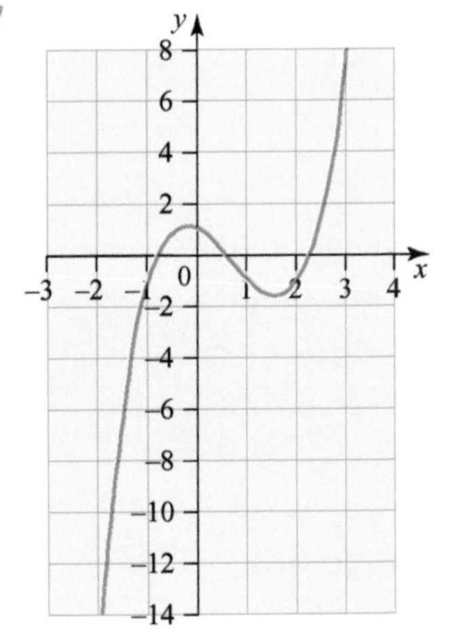

8

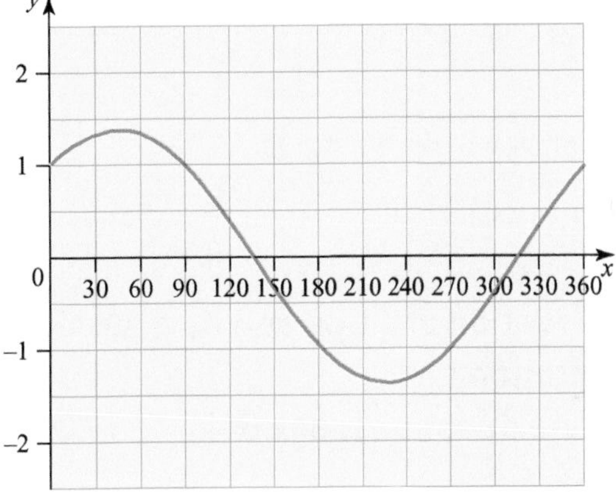

9

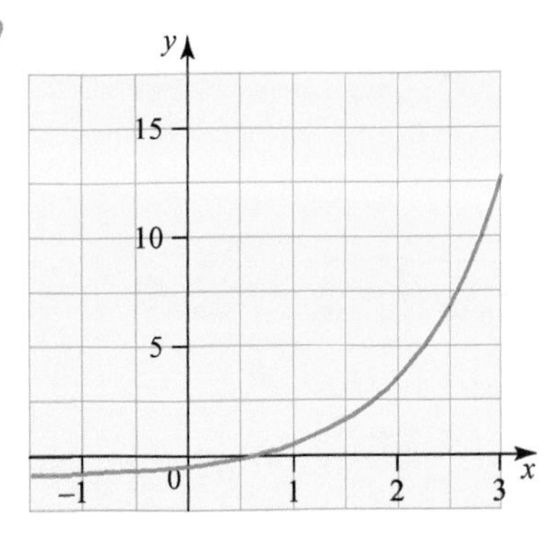

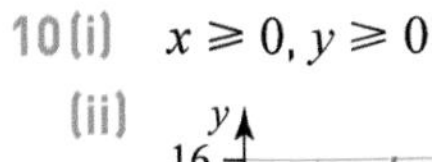

10 (i) $x \geqslant 0,\ y \geqslant 0$

(ii)

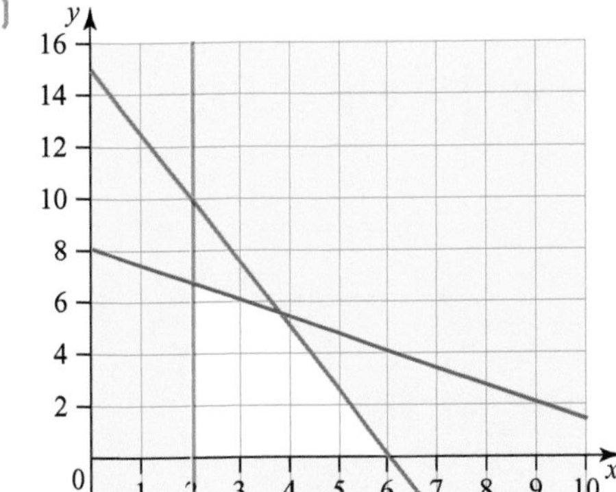

(iii) 13

Chapter 5 Points, lines and circles

5.1 Points and lines

Exam-style question (page 30)

(i) Go online for the full worked solution.

(ii) Go online for the full worked solution.

(iii) Go online for the full worked solution.

(iv) Go online for the full worked solution.

5.2 The circle

Exam-style question (page 32)

(i) $x^2 + (y-1)^2 = 9$

Any correct form of the equation would be acceptable.

(ii) $2\sqrt{8}$

(iii) The point is outside the circle.

Chapter 6 Graphs

6.1 The equation of a line and its graphical representation

Exam-style question (page 34)

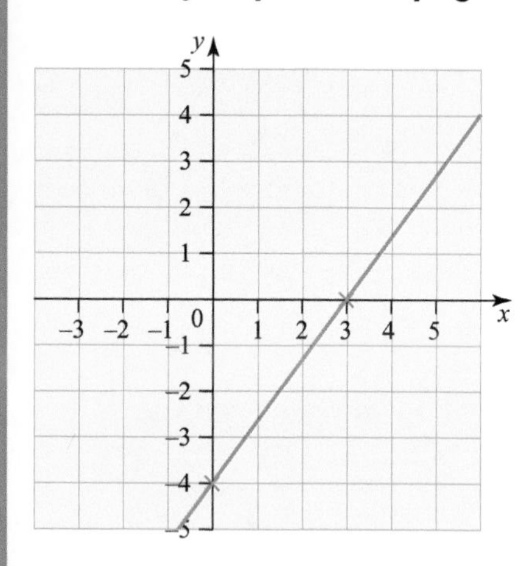

6.2 Plotting or sketching polynomial functions

Exam-style question (page 36)

(i)

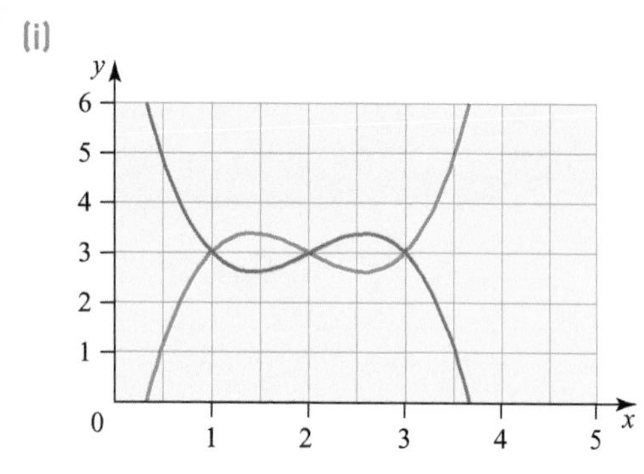

(ii) $x = 1, 2, 3$

6.3 Trigonometric and exponential functions

Exam-style question (page 39)

(i)

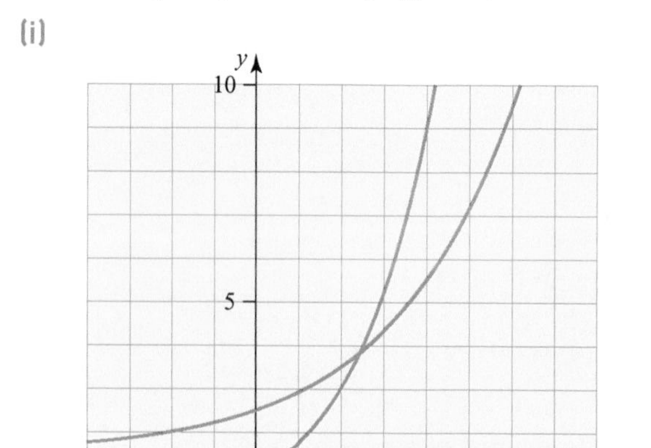

(ii) An estimate is 1.2

(iii) Go online for the full worked solution.

Chapter 7 Linear inequalities in two variables

7.1 Linear inequalities in two variables

Exam-style question (page 43)

(i) $45x + 50y \leqslant 700$

$x \geqslant 3y$

$y > 0$

(ii)

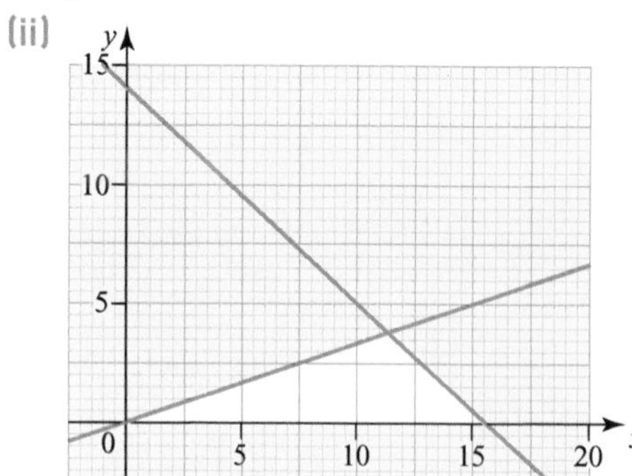

(iii) The manufacturer should make 12 chairs and 3 tables.

Review questions (Chapters 5–7) (pages 44–45)

1 $y = 9x - 12$

2 (i) Gradient of line $\text{AC} = -\frac{1}{3}$

Gradient of line $\text{BD} = 3$

The gradients satisfy the condition $m_1 m_2 = -1$ and so are perpendicular.

(ii) Go online for the full worked solution.

3 (i) Go online for the full worked solution.

(ii) $2y = x + 11$

4 (i) Centre (1,2), radius 5

(ii) $(-3, -1)$ and $(4, 6)$

5 (i) $2y = 3x + 1$

(ii) (3, 5)

(iii) (0, 7)

(iv) $x^2 + (y-7)^2 = 13$

6 (i) Centre (2, 3), radius 5

(ii) Go online for the full worked solution.

(iii) $4y + 3x = 43$

(iv) $4y + 3x = -7$

7 (i)

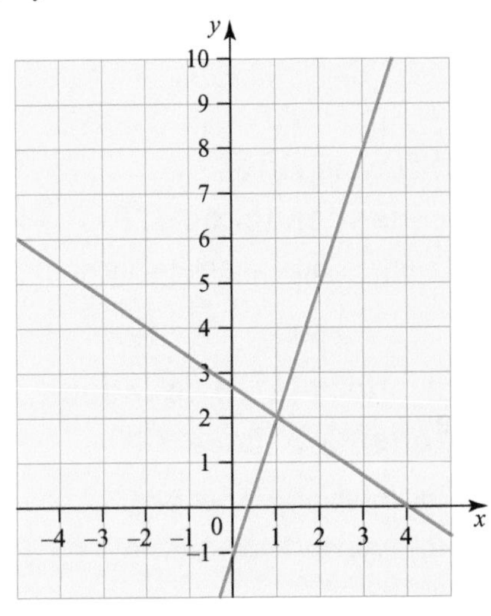

(ii) (1, 2)

8 (i) Two turning points

(ii) (–1, 8) and (3, –24)

(iii) When $x = 4$ and $x = -2$

9 (i)

θ	0	30	60	90	120	150	180
sin θ	0	0.5	0.87	1	0.87	0.5	0
cos (θ – 30)	0.87	1	0.87	0.5	0	–0.5	–0.87

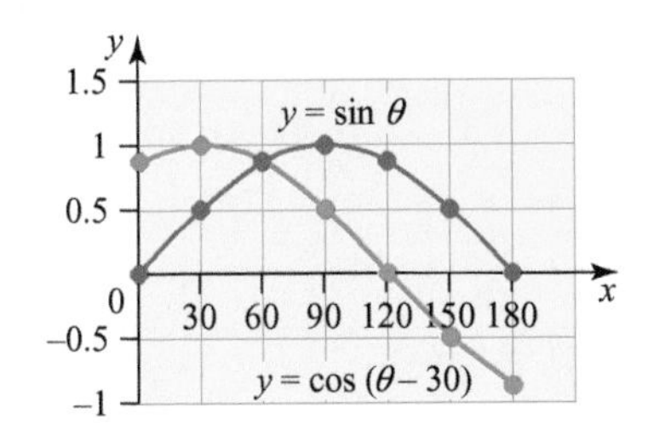

(ii) $\theta = 60°$

10 (i) Because, for any value of $a > 0$ $y = ka^x$ goes through (0, 3).

(ii)

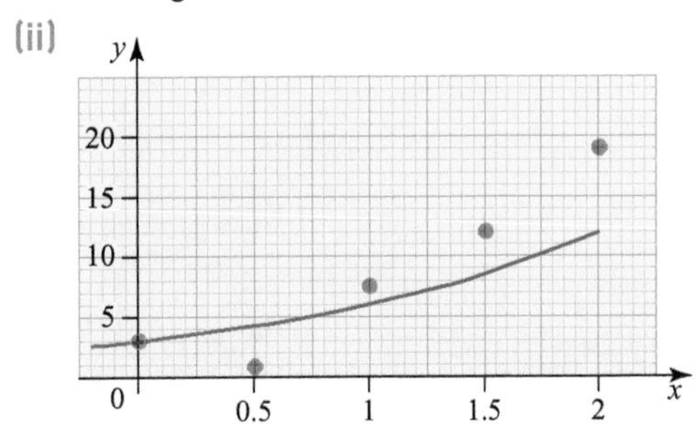

Clear indication that the data are above the curve.

(iii) $a > 2$

11 (i)

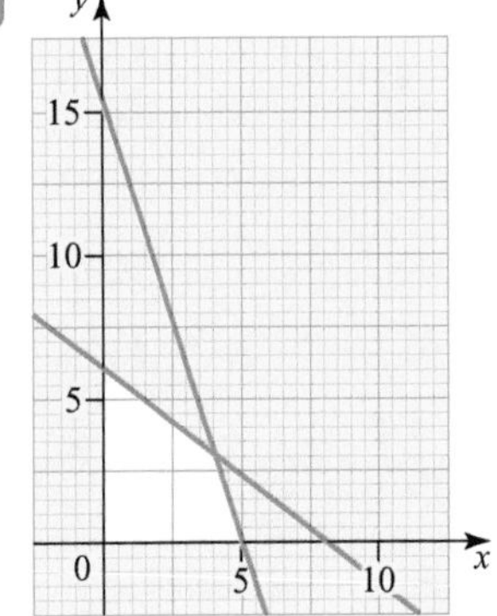

(ii) The line $x + 2y = 11$ is shown. This is the greatest value if x and y are to be integers.

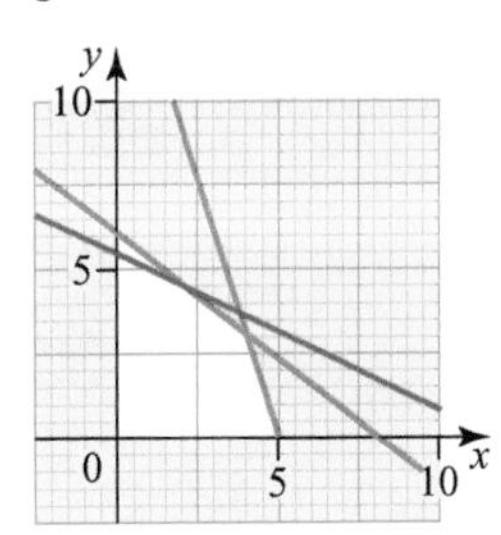

12 (i) Cost: $18x + 11y \leqslant 200$

Man hours: $7x + 6y \leqslant 84$

Customer demand: $x \geqslant 2$

and $y \geqslant 2$

(ii)

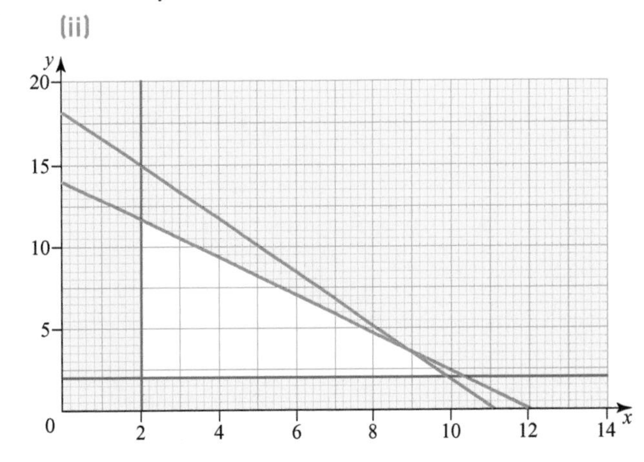

(iii)

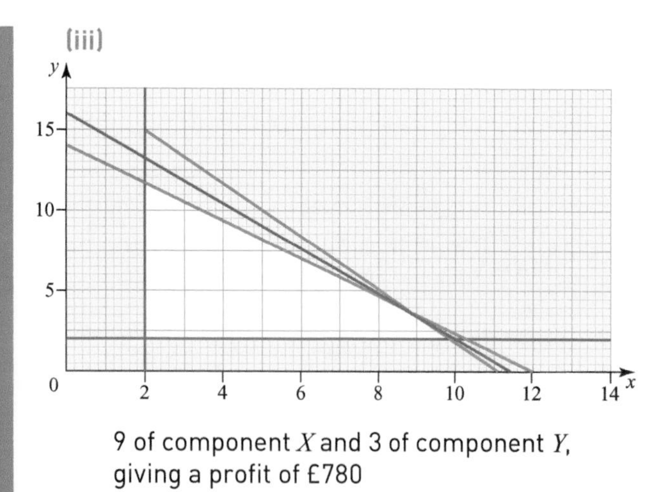

9 of component X and 3 of component Y, giving a profit of £780

Section 3 Trigonometry

Target your revision (Chapters 8–9) (page 46)

1 23.6°, 156.4° correct to 1 d.p.

2

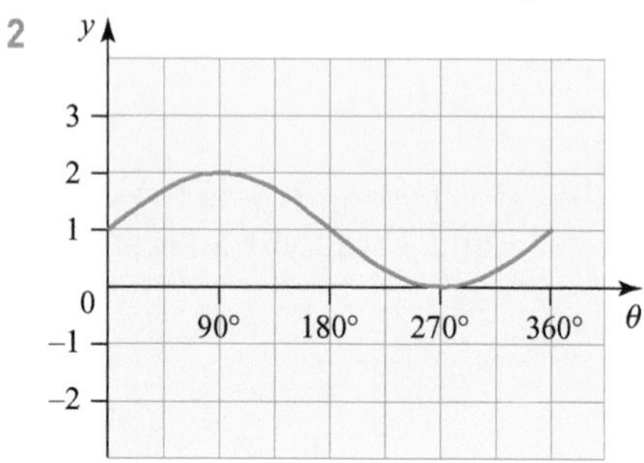

3 9.40 cm^2 to 3 s.f.

4 3.22 cm, correct to 3 s.f.

5 PQ = 7.32 cm, correct to 3 s.f.
RQ = 6.62 cm, correct to 3 s.f.

6 C = 50.5° or 129.5°, correct to 1 d.p.

7 Go online for full worked solution.

8 θ = 63.4° or 243.4°, correct to 1 d.p.

9 Angle of greatest slope = ∠ DAF = 21.8°

10 CAE = 10.1°

Chapter 8 Trigonometric functions

8.1 Angles greater than 90° Exam-style question (page 49)

(i)

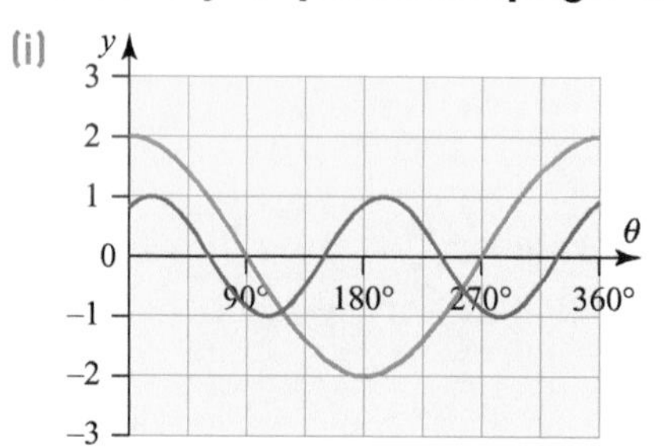

(ii) θ = 117°

8.2 Sine and cosine rules Exam-style question (page 52)

Adam and Beth are 2.1 km apart, to 2 significant figures.

8.3 Identities and related equations Exam-style question (page 53)

(i) Go online for full worked solution.

(ii) Go online for full worked solution.

(iii) 75° or 15°

Chapter 9 Applications of trigonometry

9.1 Applications of trigonometry Exam-style question (page 57)

58.8°

Review questions (Chapters 8–9) (pages 58–59)

1 x = 15°, 75°, 195°, 255°

2 216.9°

3 x = 116.6°, 296.6°

4 (i) Area = 24 cm^2
(ii) 56.4°, 123.6°

5 9.47 m

6 25.8° or 334.2°

7 Go online for full worked solution.

8 (i) 8.39 m

(ii) 9.10 m

9 (i) Area ABE = 50 000 m^2
Length BG = 200 m

(ii) 10.7°

(iii) BG ≈ 186 m

Section 4 Selections

Target your revision (Chapters 10–11) (page 60)

1

	Walked	Brought by car	Came by bus	Total
Boys	5	3	10	18
Girls	4	5	3	12
Total	9	8	13	30

2 10

3

Tennis | Badminton

win = 0.7 → win = 0.4: w, w = 0.7 × 0.4 = 0.28

win = 0.7 → lose = 0.6: w, l = 0.7 × 0.6 = 0.42

lose = 0.3 → win = 0.4: l, w = 0.3 × 0.4 = 0.12

lose = 0.3 → lose = 0.6: l, l = 0.3 × 0.6 = 0.18

Probability that Ahmed wins both games = 0.28

4 $\frac{7!}{4!} = 210$

5 (i) 120 ways

(ii) 48

6 4060

7 (i) $\frac{1}{4}$

(ii) $\frac{3}{14}$

8 $32 - 240x + 720x^2 - 1080x^3 + 810x^4 - 243x^5$

9 $1 + 5x + \frac{45}{4}x^2 + 15x^3 + \frac{105}{8}x^4 + \ldots$

10 –4320

11 (i) 0.349 (3 s.f.)

(ii) 0.264 (3 s.f.)

Chapter 10 Permutations and combinations

10.1 Definitions and probability Exam-style question (page 64)

(i)

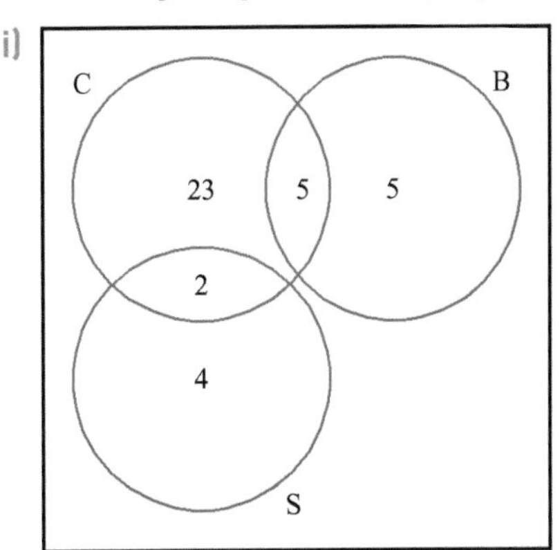

(ii) 6 people did not buy any of these three items.

10.2 Factorials, permutations and combinations Exam-style question (page 66)

(i) 720 ways

(ii) 120 ways

(iii) 48 ways

Chapter 11 The binomial distribution

11.1 The binomial expansion Exam-style question (page 68)

1 $1 - 12x + 66x^2 - 220x^3$

2 90 720

11.2 The binomial distribution Exam-style question (page 70)

(a) (i) P(all 5 red) = 0.168

(ii) P(2 red) = 0.132

(b) There are two outcomes – green or red (equivalent to 'success' and 'failure').

The probability of picking a red marble remains constant.

Picking a marble is independent of the previous selection.

Review questions (Chapters 10–11) (page 71)

1 (i)

	2nd die					
1st die		1	2	3	4	5	6
	1	3	4	5	6	7	8
	2	5	6	7	8	9	10
	3	7	8	9	10	11	12
	4	9	10	11	12	13	14
	5	11	12	13	14	15	16
	6	13	14	15	16	17	18

(ii) $\frac{1}{12}$

2 (i)

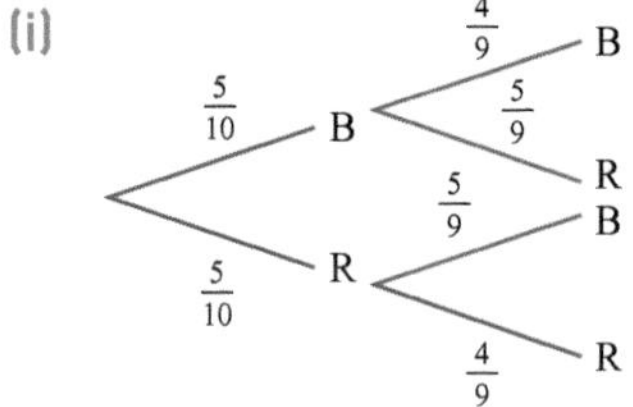

(ii) $\frac{5}{18}$

(iii) $\frac{5}{9}$

3 (i) 252

(ii) 1260

4 (i) 720

(ii) 48

5 $1024x^5 - 1280x^4 + 640x^3 - 160x^2 + 20x - 1$

6 $n = 6, \quad a = \frac{1}{2}$

7 (i) $1 + 4x + 7x^2 + 7x^3$

(ii) $\left(1 - \frac{2}{x}\right)^2 = 1 - \frac{4}{x} + \frac{4}{x^2}$

(iii) 13

8 (i) $\frac{3}{8}$

(ii) $\frac{15}{16}$

9 (i) 0.00168

(ii) 0.257

(iii) 0.917

Section 5 Powers and iteration

Target your revision (Chapters 12–13) (pages 72–73)

1

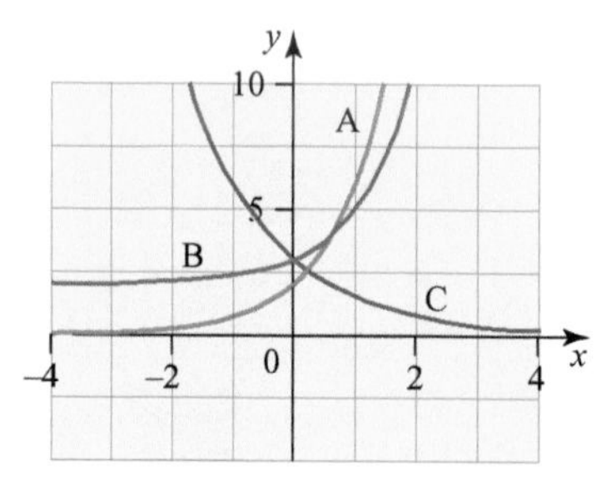

2 (i) 20 mg

(ii) 1.57 mg

3 $\log\left(\frac{4\sqrt{14}}{7}\right)$

4 (i) Go online for full worked solution.

(ii)

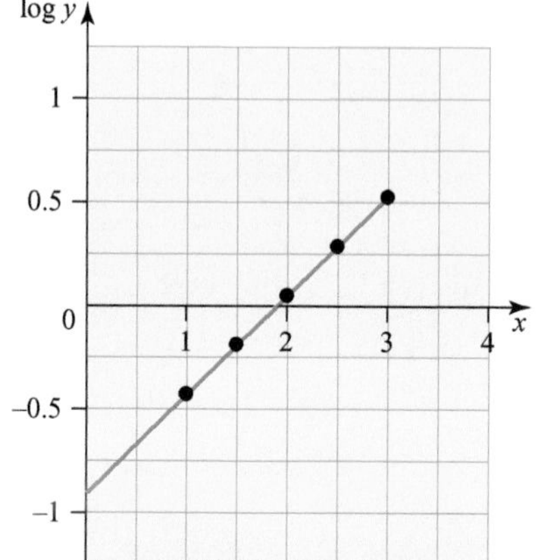

(iii) $k \approx 0.1, \ a \approx 3$

5 Go online for full worked solution.

6 $x = 2.81$, correct to 3 s.f.

7 $x = 0.377$

8 The root is 1.21 correct to 2 d.p.

9 The root is 1.77 correct to 2 d.p.

10 (i) Go online for full worked solution.

(ii) Go online for full worked solution.

(iii) The root is 1.6717, correct to 4 d.p.

11 (i)

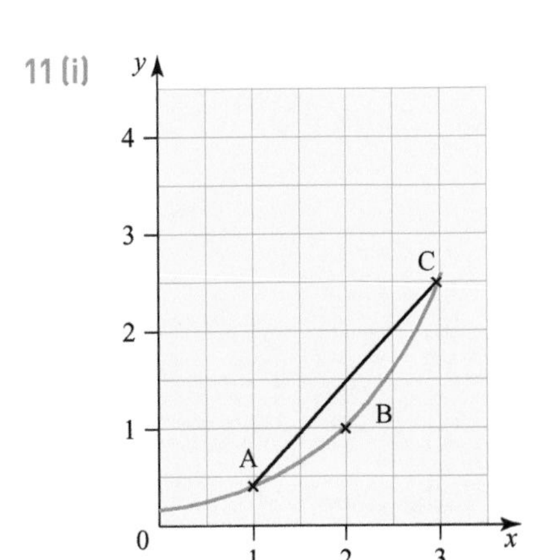

(ii) Gradient of chord AC = 1.09

Because A and C are equidistant either side of B, the chord will be approximately parallel to the tangent at B.

So the gradient of the tangent can be estimated to be 1.1.

12

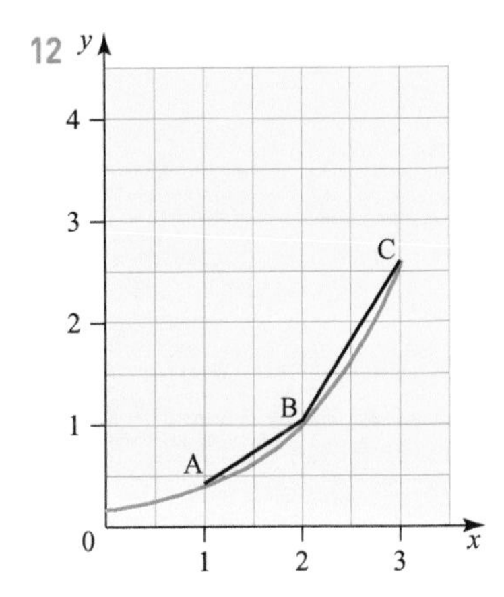

(i) Gradient of chord BC = 1.56
This is greater than the tangent at B
Gradient of chord AB = 0.62
This is less than the gradient of the tangent at B.

(ii) These values do not give an accurate answer as they are rather far apart. A better estimate is the mean of these values which is 1.09.

13 (i) Area = 7 square units.

(ii) Area = 14 square units.

(iii) A better estimate is the mean of these values. i.e. 10.5 square units

14 10.5 square units

Chapter 12 Exponentials and logarithms

12.1 Exponential functions
Exam-style question (page 76)

£3797

12.2 Logarithms
Exam-style question (page 78)

(i) $\log_{10}\left(\frac{x^2}{6}\right)$

(ii) $x = 9.75$

12.3 Reduction to linear form
Exam-style question (page 81)

(i) Go online for full worked solution.

(ii) $k = 100$

$a = 1.2$

(iii) Once the number of customers reaches the number of people in the town who buy meat, it can grow no further.

12.4 Solving exponential equations
Exam-style question (page 82)

(i) Go online for full worked solution.

(ii) Go online for full worked solution.

(iii) The temperature is approaching 25°C.

Chapter 13 Numerical methods

13.1 Sign change methods for solving an equation
Exam-style question (page 85)

(i) $f(1) = -2 < 0$, $f(2) = 6 > 0$

There is a root in the interval (1, 2)

(ii) The root is $x = 1.4$ to 2 s.f.

13.2 Iterative methods
Exam-style question (page 88)

(i) Go online for full worked solution.

(ii)

(iii)

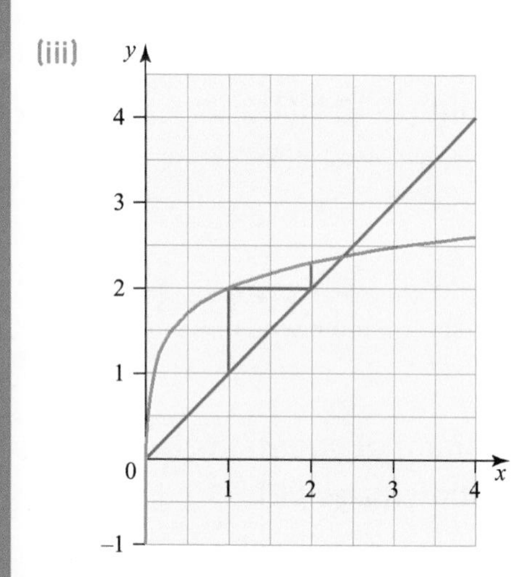

(iv) $x = 2.38$

13.3 Gradients of tangents
Exam-style question (page 89)

(i) and (ii)

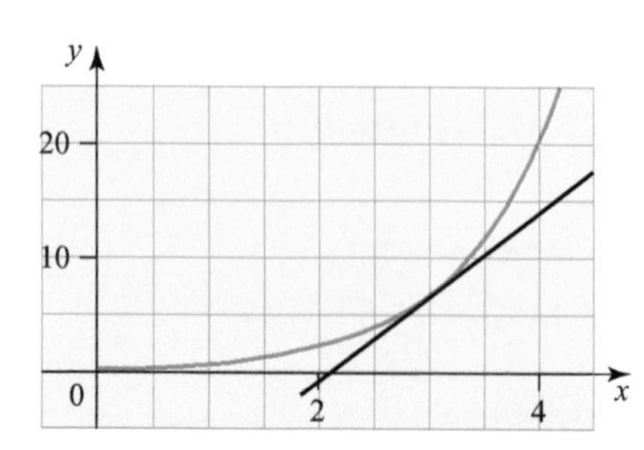

(iii) Gradient = 9

(iv) Accuracy can be improved by taking points closer to the point where $x = 3$. In this case perhaps $x = 2.5$ and $x = 3.5$.

13.4 The area under a curve
Exam-style question (page 91)

(i)

(ii) $A = 40$

(ii) The first 2 trapezia are under the curve but the third is just over. The estimate will be an underestimate.

Review questions (Chapters 12–13) (page 92)

1 $a = 3$, $k = 0.5$

2 The sum is doubled by the end of the 15th year.

3 $x = 200$

4 $3\log_2 4 - 5\log_2 8$

$= 3\log_2 2^2 - 5\log_2 2^3$

$= 6\log_2 2 - 15\log_2 2$

$= 6 - 15 = -9$

5 (i) Go online for full worked solution.

(ii)

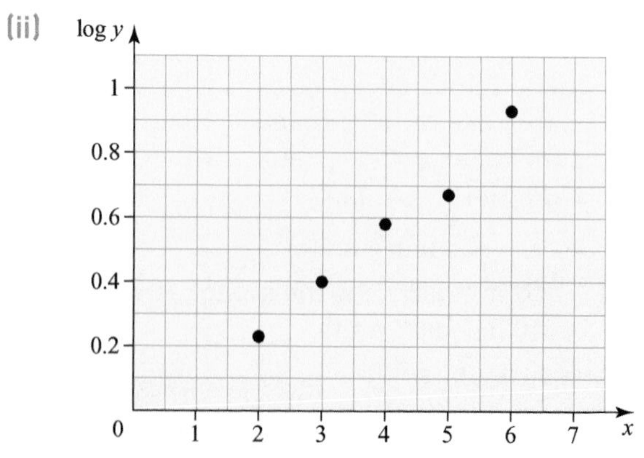

(iii) The point (5, 4.7) does not fit the model.

(iv) 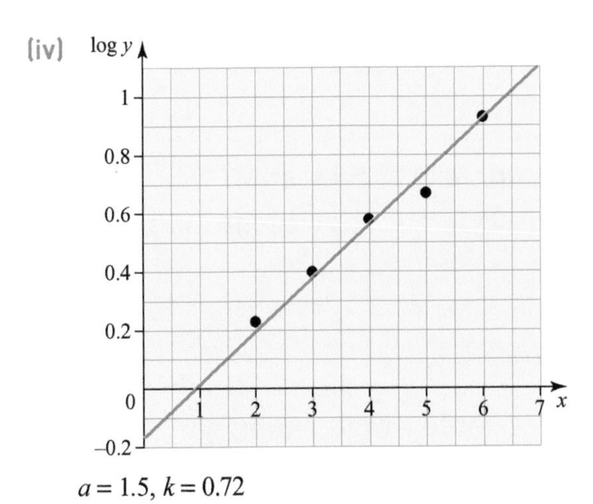

$a = 1.5, k = 0.72$

6 (i) Go online for full worked solution.

(ii) 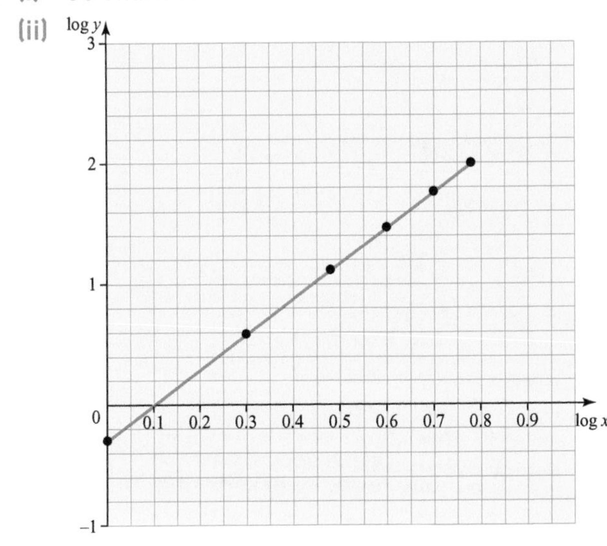

(iii) $k = 5, n = 3$

7 $x = 6.23$

8 The sum exceeds £7000 by the end of the 12th year.

Section 6 Calculus

Target your revision (Chapters 14–16) (pages 93–94)

1 $\frac{dy}{dx} = 4x + \frac{1}{2}$

2 2

3 $y = 12x - 17$

4 $12y + x = 86$

5 Maximum at (−3, 17)

Minimum at (1, −15)

6 $x^3 - \frac{3x^2}{2} + x + c$

7 $y = x + x^2 - x^4 + 15$

8 $\frac{40}{3}$

9 Area $= 23\frac{1}{3}$

10 Area $= 2\frac{2}{3}$

11 (i) $1.35\ \text{ms}^{-2}$

(ii) 11.1 s

12 (i) The straight line indicates constant acceleration.

(ii) $1.5\ \text{ms}^{-2}$

(iii) 75 m

13 (i) 16.5 m

(ii) $t = 3.7\ \text{s}$

14 (i) $a = -6\ \text{m s}^{-2}$

The particle is slowing down.

(ii) $8\frac{1}{3}$ m

(iii) 6 s

15 (i) $t = 30\ \text{s}$, $v = 27\ \text{m s}^{-1}$

(ii) 405 m

Chapter 14 Differentiation

14.1 Differentiation and gradients of curves

Exam-style question (page 97)

(i) 21

(ii) Go online for full worked solution.

14.2 Tangents and normals

Exam-style question (page 98)

(i) (1, −4) and $\left(\frac{7}{2}, -\frac{11}{4}\right)$

(ii) The line is a normal to the curve at the point (1, −4) but not at $\left(\frac{7}{2}, -\frac{11}{4}\right)$.

14.3 Stationary points and the second derivative
Exam-style question (page 101)

(i) The coordinates of the other turning point is (−1, 12)

(ii) Maximum

(iii)

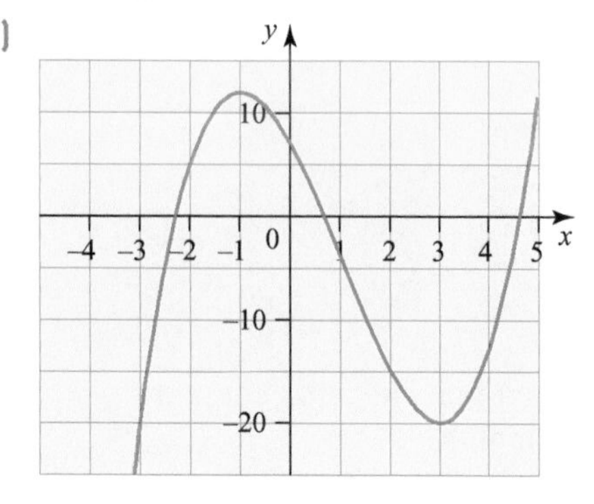

Chapter 15 Integration

15.1 Integration
Exam-style question (page 103)

(i) $x^2 + 2x + c$

(ii) $y = x^2 + 2x + c$

(iii)

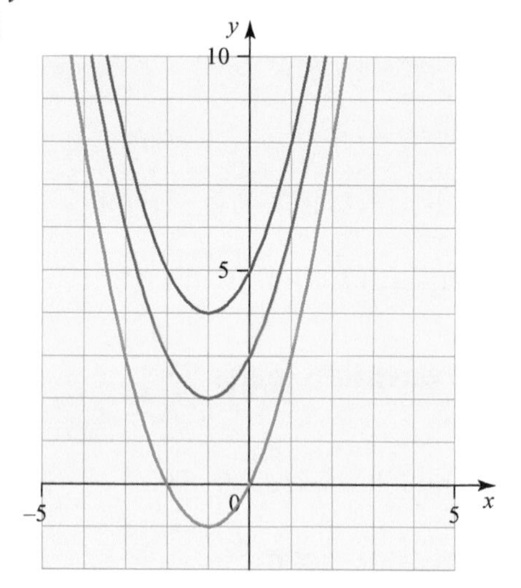

(iv) $y = x^2 + 2x - 15$

15.2 Definite integrals and area
Exam-style question (page 107)

(i) $x = 1$ or 3

(ii) $2\frac{2}{3}$ square units.

Chapter 16 Applications to kinematics

16.1 Constant acceleration
Exam-style question (page 109)

(i) $2.5\,\text{m s}^{-2}$

(ii) 80 m

(iii) 6 s

16.2 Variable acceleration
Exam-style question (page 111)

(i) $26\frac{1}{2}\,\text{m s}^{-1}$

(ii) $65\frac{5}{6}\,\text{m}$

Review questions (Chapters 14–16) (pages 112–113)

1 (3, 10)

2 $6y + x = 13$

3 (i) $y = 2x + 1$

(ii) (−4, −7)

(iii) No, this line is not a normal to the curve at R.

4 (i) Go online for full worked solution.

(ii) (3, 5)

(iii)

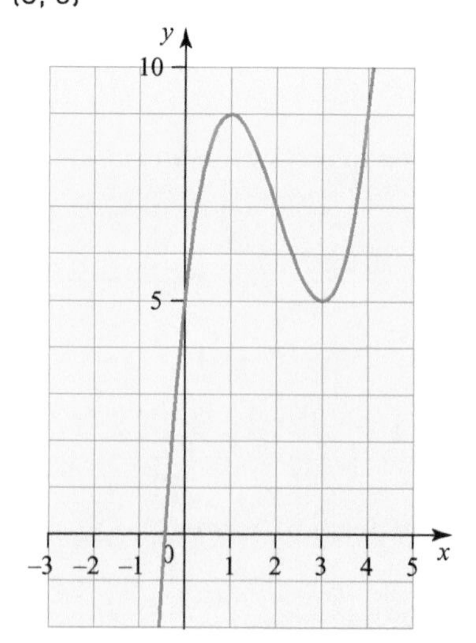

5 (i) $(-2, 14)$ and $\left(1, \frac{1}{2}\right)$

(ii) $(-2, 14)$ is a maximum

$\left(1, \frac{1}{2}\right)$ is a minimum

(iii)

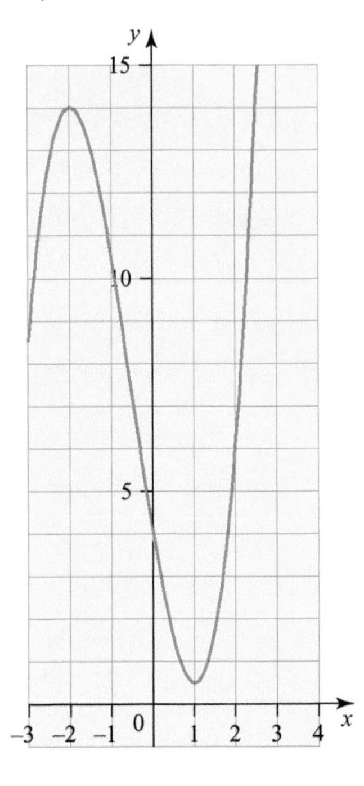

6 $y = 2x + x^2 - \frac{x^3}{3} + 4$

7 $\frac{32}{3}$

8 (i) A is at (1, 2) and B is at (7, 8)

(ii) 72

9 (i) $26\,\mathrm{m\,s^{-1}}$

(ii) 115 m

10 8 s

11 (i) $a = \frac{dv}{dt} = 0.72t - 0.072t^2$

(ii) 60 m